Otto Stephenson

Stephenson's Illustrated Practical Test,

examination and ready reference book for stationary, locomotive and

marine engineers, firemen, electricians and machinists to procure steam

engineer's license. Vol. 2

Otto Stephenson

Stephenson's Illustrated Practical Test,
examination and ready reference book for stationary, locomotive and marine engineers, firemen, electricians and machinists to procure steam engineer's license. Vol. 2

ISBN/EAN: 9783337421335

Printed in Europe, USA, Canada, Australia, Japan

Cover: Foto ©berggeist007 / pixelio.de

More available books at **www.hansebooks.com**

STEPHENSON'S

ILLUSTRATED

PRACTICAL TEST, EXAMINATION

AND

Ready Reference Book

FOR

STATIONARY LOCOMOTIVE AND MARINE ENGINEERS

...Firemen, Electricians and Machinists...

TO PROCURE

STEAM ENGINEER'S LICENSE

ALSO

WORKING CHART

For Setting Out the Forms of Gear Teeth so That Any Two Wheels of a Set May Work Together.

CHICAGO
LAIRD & LEE, PUBLISHERS
1896

AUTHOR'S PREFACE.

The reason for the publication of this book is, that having given over 25 years of my life to the careful study and practical workings of Boilers, Engines, Pumps, Electric Light Dynamos, Turning Laths, Planers, Shapers and general Machine Shop practice, I thought it proper to give the rising young mechanics and engineers such information in plain simple language so as to be easily understood, and save years of time and money in gathering the information through other sources.

It is only necessary to say that the longer one labors in practical mechanics, the more mature is his mind and judgment and the better qualified he is to carry on his work. This book is written especially for Mechanics who wish to prepare themselves in procuring Government, State or City License as an Engineer. Hoping the practical suggestions throughout the book will enable those who look and follow them up to gain a better insight of the work they have to perform,

I remain,
Very Respectfully
OTTO STEPHENSON.

PUBLISHER'S NOTICE.

We desire to place a copy of this work in the hands of every Engineer, Fireman, Machinist and Electrician, and, if the neighborhood has no agent through whom it can be purchased, we will send by mail or express, free of postage, single copies to any address on receipt of regular retail price, $1.00.

We at all times desire Agents.

The terms are liberal, and the agency to sell this work in any field will afford a good living to any man or woman of intelligence.

Address all communications to

LAIRD & LEE
Publishers,
**263-265 WABASH AVENUE
CHICAGO.**

STEPHENSON'S PRACTICAL TEST,
—FOR—
Steam Engineers, Firemen, Electricians and Machinists.

Ques. Name the duties of an Engineer. **Ans.** The duties are to take full charge of the boilers and engines where ever he may be employed, and see that the steam machinery under his charge are kept in No. 1 order.

Ques. What is required of a man to become a Chief Engineer? Ans. He is obliged to obtain an Engineer's license.

Ques. What experience should a man have in order to get his application before a board of Engineers? Ans. The experience should be generally two years, as a Fireman, Machinist or Engineer which must be sworn to by two good reliable citizens, both living in the city where the applicant has been employed.

Ques. What are steam boilers? how are they, and of

what are they made? Ans. Steam boilers are closed vessels made of steel, iron or copper plates, the most plates in use are $\frac{3}{8}$, $\frac{1}{2}$ and $\frac{5}{16}$ inch, and the tensile strength ranging from 45,000 to 85,000 lbs. These plates are run through a rolling machine and rolled in a circle, then the sheets are riveted together at the end with two rows of rivets.

Ques. Why? Ans. Because the strain is greater sidewise than endwise, the seams around the boiler are single riveted because the strain is not so great.

Ques. Why is the strain greater on the sides than on the ends of the boilers? Ans. Because the steam has more surface on the sides, and only the heads on the end of the boiler to contend with.

Ques. How is a boiler strengthened? Ans. With braces.

Ques. Name one of the braces, and how are they put in? Ans. The boiler is braced by different kinds of braces, such as a crow-foot, longitudinal, dome, crown-bars, angle braces, etc. The eye is riveted to the head of the boiler, which head is generally made of $\frac{5}{8}$ inch plate, the other eye is riveted to the side top or dome of the boiler, and the braces and eyes are put together with bolts, which have a split-key to keep the bolt in its place.

Ques. Name the chief points in the construction of a successful and economical boiler? Ans. Proper circulation facilities constitute one of the chief points in the construction of a successful and economical boiler. In tubular boilers, the best practice is to place the tubes in vertical rows, (plumb) leaving out what would be the centre row. The circulation is up the sides of the boiler and down the centre. Tubes set zigzag, or to break spaces, check the circulation and will not practically give the best results.

Ques. How should braces fit? Ans. They should fit tight, otherwise they would be of no use.

Ques. If braces were found loose, what should be done and how could they be tightened? Ans. The braces should be taken out heated in the centre, then upset by dropping it endwise on a block of wood until it is the proper length.

Ques. State the strain on a brace? Ans. One fifth of its own strength.

Ques. What are stay bolts? Ans. They are long screw bolts of one continuous thread.

Ques. Where are they generally used? and name the reasons for using them. Ans. They are generally used for keeping two sheets apart in order to leave a water space between them, such as a locomotive fire box and shell.

Ques. How should stay bolts be spaced and how far apart? Ans. The surface which a stay-bolt has to support is represented by the rectangle enclosed between four of the bolts. For instance, if the stays are six inches apart, the area supported by each would be 6x6—36 sq. inches. Multiplying this area by the boiler pressure gives the stress upon each stay, and the stays should be set at such distances from each other that the stress shall not exceed 6,000 pounds per square inch of cross sectional area. To determine this distance multiply the cross sectional area of the bolt by 6,000; divide by the boiler pressure, and extract the square root of the quotient.

Ques. What is meant by the term corrosion? Ans. It means the wasting away of the boiler plates by pitting, grooving, etc.

Ques. Name the different corrosions? Ans. There are internal and external corrosions.

Ques. Explain their actions? Ans. The acids and

minerals in the water liberated by the heat, causes the internal, and the sulphur which is liberated from the coal by fire causes the external.

Ques. How is the water level found when a boiler is foaming? Ans. The proper way would be to shut down the engine and all valves connected with the boiler, cover the fire with ashes and close the damper, then the water will quiet down, and the water level easily found. An Engineer should know when lighting a fresh fire, never to force it, but let it burn slowly so that all the parts will expand as near equal as possible; good judgment is needed. Test the boilers and steam guages at least once a year.

Ques. Where is a steam guage generally tapped in a boiler? Ans. On top of the boiler in the steam drum. It must always be tapped into the steam part of the boiler.

Ques. With what should the steam gauge correspond? Ans. The steam gauge should correspond by all means with the safety valve.

Ques. Why is a drain cock put under the steam gauge? Ans. To drain the pipe in cold weather.

Ques. Why are steam gauges used? Ans. They are used to indicate the pounds pressure per square inch in the boiler.

Ques. Do steam gauges get out of order? Ans. Yes, sometimes.

Ques. Suppose the steam gauge was wrong what would you be governed by? Ans. By the safety valve only.

Ques. How would you know the safety valve was in working order? Ans. By raising the lever two or three times carefully to see that the valve worked free and is not stuck.

Ques. Of what use is a safety valve? Ans. It is supposed to release the boiler from every pressure of steam.

Ques. What size should the safety valve be in proportion to the grate surface? Ans. The safety valve should be about ½ square inch to each square foot of grate surface, which makes it large enough to relieve the boiler of all steam over which the safety valve is set.

Ques. Which are the better, gauge cocks or glass gauges, and which is most preferable? Ans. Gauge cocks, because the glass gauge is liable to get choked with mud, and not give a true level of the water, the glass gauge is a very handy thing; it should be blown out four or five times a day, so as to keep it free from mud.

Ques. What should be done in case a glass should break? Ans. Close the water valve first to prevent the escape of water, then close the steam valve. Insert a new glass, then turn on the steam valve first, the water valve next, then close the pet cock at the bottom.

Ques. Which is the better way to clean the inside of a glass gauge? Ans. Tie a small piece of waste to a strong thin stick, saturate with soap or acetic acid, pass down inside of the glass, then open the steam valve and blow steam through the glass, and the glass will be clean. Never touch the inside of a glass with a wire, as it will crack.

Ques. If a gauge cock, or a small steam pipe in the large steam pipe should happen to break off, what should be done? Ans. Simply make a hard wood plug and drive it in with a heavy hammer, which should be left so until the break could be repaired.

Ques. How is it repaired? Ans. By cutting out the old piece, retapping and putting in another pipe or gauge cock, whichever the case may be.

Ques. What side clearance should boilers have between the furnace walls and shell at the fire line? They should have from three to four inches at the fire line, and from seven to ten inches between the shell and the bridge wall. The space where the smoke returns to pass through the flues should be larger than the area combined of the flues or tubes, the bridge wall should pitch toward the back.

Ques. How should stationary boilers rest and what on? Ans. The front end of the boiler should rest on the fire front, and the back end on a cast iron stand or saddle to allow equal expansion, the mud drum should always hang free under all circumstances. (If the boiler has one).

Ques. In what should Engineers be careful and exercise good judgment? Ans. In starting or stopping an engine with a high pressure of steam.

Ques. Why should engineers be careful in starting or stopping an engine? Ans Because the vent given the steam in starting, and the sudden check in stopping may cause such a pressure as to rupture the boiler.

Ques. How large a hole would it be safe to cut in an iron boiler five-sixteenths of an inch thick, without putting a flange around it? Ans. In practice, a two-inch pipe is often put in without giving any trouble, but is usually done when the boilers are located at a distance from the shop, or the time to put on a stiffening piece can not be spared. The stiffening piece should always be put on when possible, and it can be made thicker than the shell, and therefore giving a better holding surface to the pipe and it also leaves a cavity or space around the pipe, which, when the pipe is used as a blow-off, allows deposits to settle into it, and find their way out when the valve is open. There is also in this case, less danger of the end of the pipe extending through the shell and preventing the mud from entering the pipe.

Ques. What else should engineers look after? Ans. Engineers should see that the draft is not choked by ashes under the boiler back of the bridge walls, and that the outside of the boiler and inside of the flues are kept clean from soot, then there will be no trouble in keeping up steam.

Ques. How are the flues or tubes of a steam boiler kept clean? Ans. By either blowing steam through them or using a flue brush.

Ques. How are flues or tubes cleaned by steam? Ans. By having a hose attached to the front end leading from the steam drum, so that the flues or tubes can be blown out from the front end. (Cleaning by the brush is the better and more popular way.)

Ques. How often should the tubes or flues be cleaned? Ans. Once a day, in the afternoon, sometimes in the morning after raising steam, according to the coal used.

Ques. Name the different strains of a boiler? Ans. To the flues or tubes it has a crushing strain, to the shell a tearing strain.

Ques. What mainly causes boiler explosions? Ans. There are various causes, such as low water, over-pressure of steam, bad safety valve, foaming boilers and burnt sheets.

Ques. Why would foaming cause an explosion? Ans.

It generally raises the water from the heated sheets. The sheets become hot; and the water falling back on them they crack, and sometimes cause an explosion. A blistered sheet or a scaly boiler will also cause an explosion, by allowing the sheets to become burnt and weakened; also an untrue steam gauge or bad safety valve is very dangerous.

Ques. Name the worst explosions? Ans. The worst explosion known, is caused by high steam pressure.

Ques. How are boilers tested for blistered, cracked or rotten sheets? Ans. By the hammer.

Ques. How is it done? Ans. By taking a small hammer and going inside and outside of the boiler and sounding it.

Ques. Explain how you would know by the sound? Ans. By the different sounds; if the sheets rings and sounds solid, they are all right; but if they sound dead, hollow or blunt, they should be condemned.

Ques. Should the iron be struck hard? Ans. Yes, pretty hard.

Ques. Is it proper to have a boiler insured? Ans. Yes as insurance is generally accompanied by the hammer test and intelligent inspection, which guarantees security to the engineer or owner.

Do not reject the advice or suggestions of intelligent boiler inspectors, as their experience enables them to discriminate in cases which never come under the observation of men who do not follow inspection as a business,

Ques. Are boilers injured by the hydraulic test. Ans. Yes, if tested by an inexperienced person.

Never use steam pressure under any circumstances for testing purposes.

Ques. If a patch is to be put on a boiler what kind

would you put on? Ans. A hard patch; it is reliable and safe.

Ques. Why not put on a soft patch? Ans. Because they are not reliable and are dangerous.

Ques. What is a hard patch? Ans. A hard patch is a patch where the piece is cut out of the boiler and rivet holes are drilled or punched through, then the patch is riveted on, chipped, caulked and made water and steam tight.

Ques. What is a soft patch? Ans. A soft patch is put over the plate that needs patching, and put on with $5/8$ or $3/4$ inch countersunk screw bolts, with a mixture of red lead and iron borings between the patch and the boiler plate; the piece of sheet in the boiler is not cut out for a soft patch as for a hard patch, consequently the soft patch is burnt.

Ques. Which are the better, drilled or punched holes? Ans. Drilled holes.

Ques. Why are the drilled holes better? Ans. Because the fiber of the iron is not disturbed as when the holes are punched; in drilling, the iron is cut out regular; in punching, it is forced out at once.

Ques. Name the proper rivets for certain sized sheets, and how far apart? Ans. The rivets should be $5/8$ to $3/4$ inch diameter, and from 2 to $2 1/4$ inches apart.

Ques. What would you do the first thing in the morning on entering the boiler-room? Ans. See how much water there is in the boiler, by trying the gauge cocks, then open the glass gauge valves, and start the fire to raise steam.

Ques. Why do you try the gauge cocks, and not trust to the glass gauge? Ans. Because the water pipe connecting the glass gauge with the boiler is liable to choke up with mud, therefore the glass would not show a true level of water. The glass gauge should be blown out eight or ten times a day, to insure safety, but never depend on the glass gauge alone.

Ques. If too much water was pumped in the boiler during the day, what should be done? Ans. Open the blow off valve and let out the water to the second gauge. An engineer should be very careful when blowing out water when there is a hot fire in the boiler furnace, as the water leaves very fast, and may blow out too much; good judgment should be used.

Ques. How is a two-flue boiler cleaned? Ans. First see that there is no fire under the boiler, then let out all the water through the blow-off valve, take out the man, hand, and mud-drum plates; then take a short-handle broom, a candle or torch, a small hand-pick, a scraper made out of an old file flattened on the end and bent to suit, also a half-inch square iron twisted link chain about three feet long, with a ring at each end to answer for a handle; place the chain around the flue and work the chain to get the scale off the bottom of the flues; use the pick and scraper to pick and scrape off all that can be seen on top of the flues and the bottom and sides of the shell; then wash out into the mud-drum; clean out and put in the mud-drum and hand-hole plates; fill up to top of flues; then put in the man-hole plate, and fill up to the second gauge ready for raising steam.

Ques. Could a boiler not be blown out? Ans. Yes, but not practically.

Ques. How much pressure would you allow? Ans. About 15 or 30 pounds.

Ques. Why not more pressure? Because the heat would be so great that the expansion and contraction between the boiler and furnace would not be equal; viz. the boiler seams would leak and the boiler injured. The practical way is no steam pressure.

Ques. What benefit is gained by letting the water stay in the boiler until ready to clean it out? Ans. The mud is kept soft and the scale is not caked to the shell or tubes; also, the seams of the boiler are not injured by unequal expansion and contraction.

Ques. How should man and hand-hold plates be taken out and put in? Ans. They should be marked with a chisel at the top, also the boiler at man hole and hand-hole, whichever it might be, and they should be put in the same way they came out.

Ques. How would you gasket the man-hole or hand-hole plates of a boiler? Ans. With pure lead rings; some use sheet rubber, etc.

Ques. Why are man-hole and hand-hole plates made oval instead of round? Ans. The practical reason is, if the holes were round the plates could not be taken out or put in, also a man could not easily enter the boiler through the man-hole.

Ques. When filling a boiler with cold water, and raising steam, what should be looked to? Ans. See that a valve is left open above the water.

Ques. Why should a valve be left open? Ans. Because boilers fill easier and quicker, and in raising steam the cold air is let out and allows equal expansion, as cold air does not allow equal expansion.

Ques. How is a boiler set? Ans. By leveling across and along the flues or tubes, allowing the end of the boiler furthest from the gauge cocks ½ inch lower for every eight feet in length.

Ques. Why is it lower? Ans. Because when there is water in the gauge cocks, there will surely be water at the other end of the boiler.

Ques. How many gauge cocks has a boiler? Ans. Generally three.

Ques. Where is the first or lower gauge? Ans. Two inches above the flues, and the rest two inches apart.

Ques. Where is the waterline? Ans. First gauge.

Ques. Where should water be in the boiler when running? Ans. Second gauge.

Ques. What should be done preparatory t shutting down for the noon hour? Ans. Slacken the draft, and let the steam run down, say 15 lbs or so, also let the water run down to below 2d gauge and when the engine has stopped put the feed on and clear the fires.

Ques. What other duties devolve on the engineer at this time? Ans. He should examine all the journals and moving parts, refill the oil cups, etc.

Ques. Where should water be carried when shutting down at night? Ans. At the third gauge and close gauge glass cocks.

Ques. Why carry water so high and close glass gauge cocks? Ans. To allow for evaporation and leakage and keep water in the boiler in case glass would break.

Ques. Where is the fire line of a boiler? Ans. ¾ of an inch below first gauge.

Ques. When you open a boiler and look in, where do the scales lay thickest? Ans. Over the fire-plates. (The second sheet generally.)

Ques. What causes that? Ans. The circulation and heat is greatest there.

Ques. Of what use is a steam drum? Ans. To have more dryer steam in volume.

Ques. How should the circulation and feed be? Ans. The circulation and feed should be continual.

Ques. Why so? Ans. Because boilers are known to have exploded immediately on the opening of the steam valve to start the engine, after the circulation in the boiler and the engine had been standing still for a short time.

Ques. Explain the cause of it. Ans. It is caused by the plates next to the fire box being overheated, and as soon as the valve is opened the pressure is lessened, and the water on the overheated sheets flash into steam and if the boiler is not strong enough, a terrific explosion is the result.

Ques. If the gauge cocks were tried and there was no water in sight what should be done? Ans. Cover the fire with wet ashes, pull the fire out, then raise the flue caps and let the boiler cool down.

Ques. Why are wet ashes thrown over the fire before pulling the fire out? Ans. To lessen the heat. If the fire was stirred up without throwing wet ashes over it, it would create more heat and very likely burn the plates.

Ques. What should be examined in the boiler every cleaning out day? Ans. The braces in the boiler should be examined to see if they are loose, also the sheets, flues, heads and seams, to see if they are cracked or leaking; if they are not attended to, they may cause serious trouble and loss of life.

Ques. What should engineers look after in and about the engine and boiler room? Ans. They should see that everything about the engine or boiler room is clean and all the tools are in their proper places. Also see that all valves or cocks do not leak, if so they should be ground in with emery and oil until a seat or true bearing is found. Ground glass is good for grinding brass valves.

Ques. When should the boiler seams be looked after and caulked? Ans. The boiler seams should be looked after when the boiler is hot, and filled with water, to find the leaks, and the caulking should be done when the boiler is cold and empty as the jarring while caulking would have a tendency to spring a leak somewhere else, if the boiler was under pressure.

Ques. Is pressure and weight the same? Ans. No.

Ques. Why? Ans. Because pressure forces in every direction, while weight presses only down.

Ques. Which is best, the riveted or the lap-welded flues? Ans. The lap-welded flues, as they are a true circle and not so easily collapsed as the riveted flues, which are not a true circle.

Ques. What is meant by foaming? Ans. Foaming is the water and steam being mixed together.

Ques. State the general causes of foaming? Ans. Dirty greasy, oily and soapy water; salt water forced into fresh water, also too much water and not enough steam room will cause foaming.

Ques. What is meant by priming? Ans. Priming is the lifting of water with steam, such as opening a valve suddenly, and drawing the water from the boiler to the ylinder of the engine.

Ques. What should be done in that case? Ans. Close the throttle valve and leave it closed for a few minutes, then open the valve slowly; that will generally remedy it. Sometimes priming is caused by too much water and not enough steam room; in that case less water is carried.

Ques. Suppose there was a high pressure of steam in the boiler and the water was out of sight, would it be safe to raise the safety valve to let off the pressure? Ans. No, under no circumstances.

Ques. Why not? Ans Because it would cause the water to rise, and when the valve closed the water would drop back on the heated parts and be liable to explode the boiler.

Ques. Suppose the boiler was too small to keep up the required amount of steam, would it be practical to weigh down the safety valve to carry a higher pressure? Ans. No, under no circumstances.

Ques. Why not? Ans. Because it would show carelessness and a violation of the laws.

Ques. Is there any mystery about boiler explosions? Ans. No, they are simply caused by carelessness. No man has the right to endanger the lives and property of others when he knows that he is incompetent to perform the duty required of him as an engineer, whether licensed or otherwise.

Ques. How much space should there be between the tubes of a steam boiler? Ans. One-half the diameter of the tube itself.

Ques. Name the principal valve on a steam boiler? Ans. The safety valve, by all means.

Ques. Where should the lower gauge cock be placed in upright boilers, any size? Ans. One-third the distance from the top, between the two flue sheets.

Ques. How long a time is it considered safe to leave the engine or boiler room alone without attention? Ans. Under no circumstances should the engine or boiler room be left alone.

Ques. Why not, when everything is in working order? Ans. Because no man can tell at what moment an accident might occur, which if neglected might cause a serious loss of life and property.

Ques. State the boiling point of water? Ans. It is 212 degrees of heat.

Ques. At what point does water evaporate into steam? Ans. It evaporates at 213 degress. (Fahrenheit.)

PUMPS.

Ques. Name the different pumps for feeding boilers? Ans. There are many kinds, but we consider only single action, double action and duplex pumps for feeding boilers and general use. (See Illustration.)

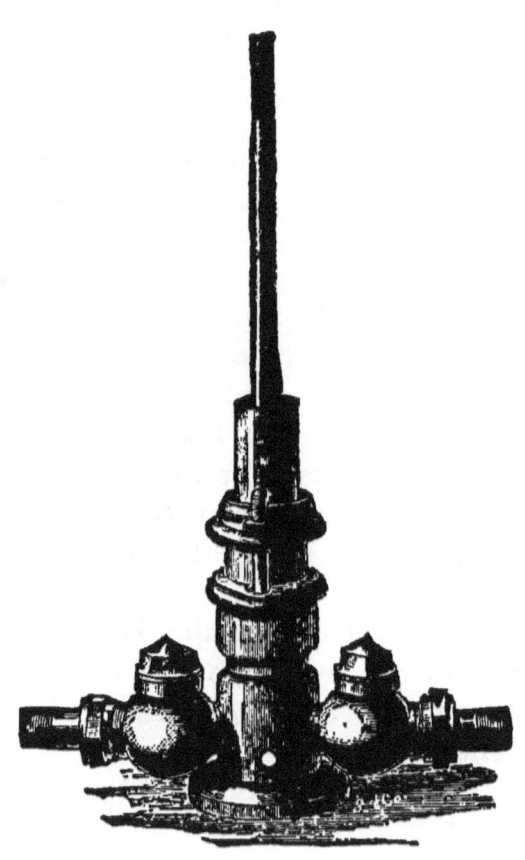

COMMON PLUNGER PUMP.

Ques. How many valves has a single action plunger pump? Ans. Two valves, a receiving valve and a discharge valve.

21

STEAM PUMP IN DETAIL.

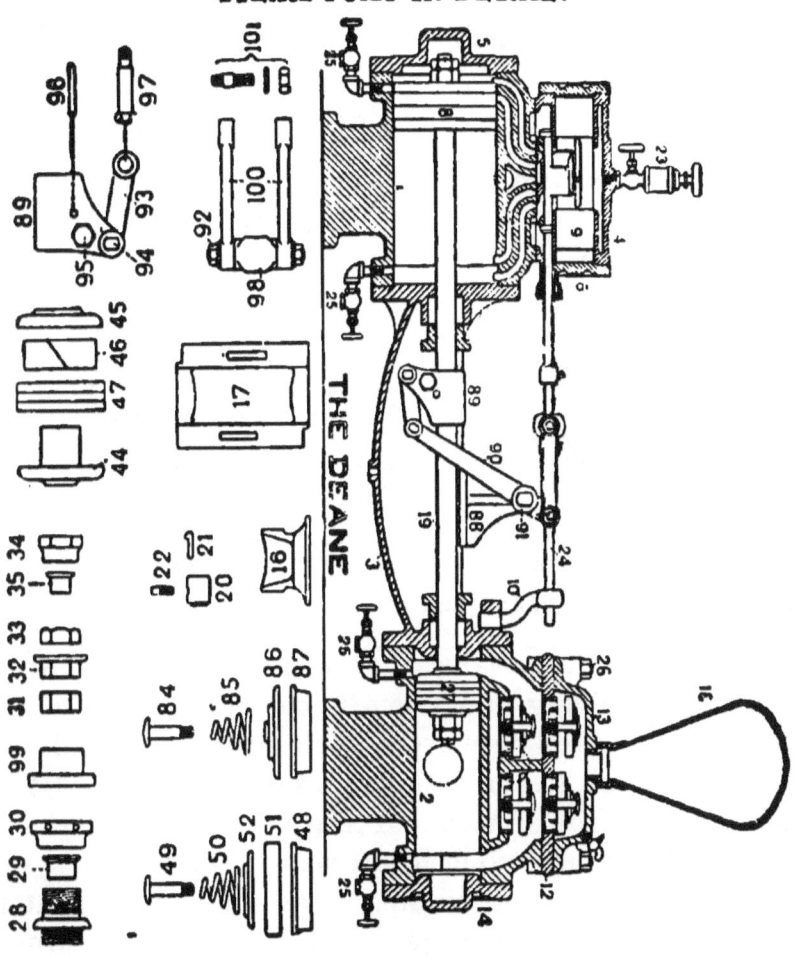

LIST OF PARTS.—NEW LEVER MOTION PATTERN.

1 Steam Cylinder,
2 Water Cylinder.

34 Cap, } For Valve Rod Stuff-
35 Gland, }   ing Box.

3 Yoke,
4 Valve Chest,
5 Steam Cylinder Head,
6 Inside Valve Chest Head,
7 Outside Valve Chest Head,
8 Steam Piston,
9 Valve Piston,
10 Guide,
12 Water Valve Plate,
13 Water Cap,
14 Water Cylinder Head,
15 Water Cylinder Lining,
16 Main Valve,
17 Auxiliary Valve,
18 Air Chamber,
19 Piston Rod,
20 Tappit,
21 Tappit Key,
22 Tappit Set Screw,
23 Lubricator,
24 Valve Rod,
25-25-25 25 Drip Plugs or Cocks,
26 Eye Bolt and Nut,
27 Water Piston,
28 Bushing, ⎫
29 Gland, ⎬ For Piston Rod stuff-
30 Cap, ⎭ ing Box.
31 Nut,
32 Flange Nut, ⎱ For Piston Rod
33 Check Nut, ⎰

44 Water Piston, ⎫
45 Follower, ⎬ For Pat. Fi-
46 Inside Ring, etc., ⎬ brous Ring
47 Fibrous Packing, ⎭ Packing.
48 Seat,
49 Stem ⎫ For Rubber Water
50 Spring, ⎬ Valve.
52 Cover, ⎭
51 Rubber Water Valve,
84 Stem, ⎫
85 Spring, ⎬ For Metal Disc Valve
87 Seat, ⎭
86 Metal Disc Valve,
88 Bearing Stand,
89 Piston Rod Arm,
90 Lever,
91 Fulcrum Pin,
92 Tappit Block Nut,
93 Piston Rod Link,
94 Link Pin,
95 Piston Rod Arm Bolt,
96 Piston Rod Arm Pin,
97 Lever Pin,
98 Tappit Block,
99 Gland for Stud Stuffing Box.
100 Valve Rod Links,
101 Link Stud, Washer and Nut.

Ques. How many valves has a double action? Ans. Four, two receiving and two discharging. The double action receives and discharges both strokes. This kind of a pump has a steam cylinder on one end. The large pumps have eight, sixteen and thirty-two small valves on water cylinder, according to the size of the pump.

Ques. Why do large pumps have many small water valves and not a few larger ones in proportion? Ans. The reason the pumps have small valves is that the valves do not have to open as much as larger ones, consequently the pump does not loose the quantity of water each stroke as it would with larger valves.

Ques. How are pumps set up and leveled? Ans. Pumps are set so the receiving is from the boiler and the discharge toward the boiler, put in the same size receiving and discharge pipe as tapped in the pump, so the pump can have a good supply and discharge. The suction should be straight as possible and perfectly air tight. The pump is leveled with a spirit-level or a square and plumb line. To level a double action pump, some level across the frame and along the piston; the other way is to take the valve chamber cap off the water cylinder and level the valve seats, so the valves will rise and drop plumb. To level a single action pump, take off the valve chamber caps and level both ways.

Ques. How are the water piston heads packed, and with what in the water cylinders? Ans. They are generally packed with square canvas packing and generally takes two or three pieces; one piece is jointed on top, and the others about $\frac{1}{3}$ way around to make, what engineers call, a broken joint. The packing runs from $\frac{1}{4}$ to $\frac{7}{8}$ inch square. These are the general sizes used for common sized pumps.

Ques. How are the steam valves of duplex pumps set and adjusted? Ans. Take off the valve chest cover, shove the piston against one of the cylinder heads and mark the piston rod with a pencil at the packing-box gland, then shove the piston against the other cylinder-head and make another mark, find the centre between the two marks and move the piston until the centre mark reaches the packing box gland where the first mark was made. Or in other words plumb the lever that connects the valve rocker shaft and the piston. After this is done, see how the steam valve is for lead; if equal at both ends the valve is set, if not, adjust by uncoupling the valve stem at the coupling outside of the packing box, and turn to suit the adjustment in equalizing the "lead."

Ques. What other valve has a pump near the boiler? Ans. A check valve.

Ques. Of what use is a check valve? Ans. To hold the

water that is forced into the boiler from coming back, in case there is any work to be done on the pump itself.

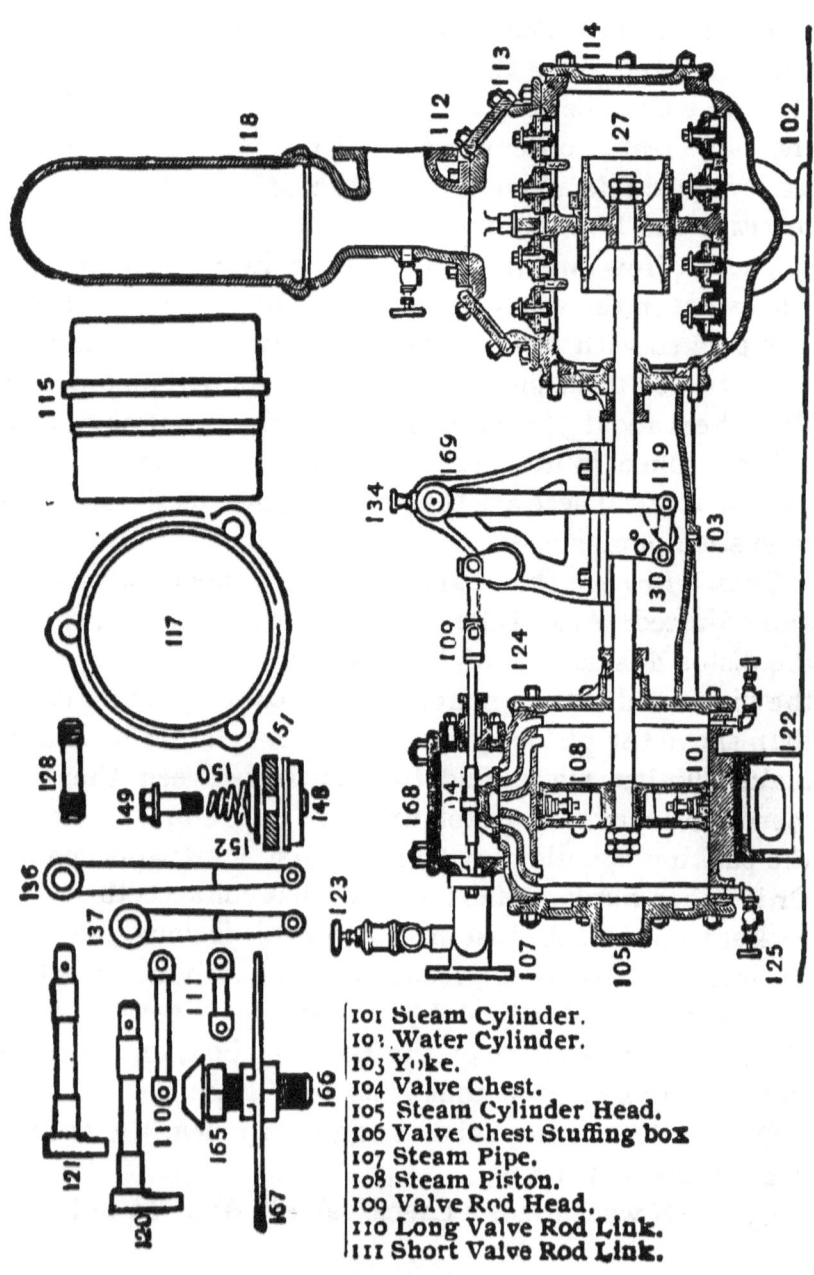

101 Steam Cylinder.
102 Water Cylinder.
103 Yoke.
104 Valve Chest.
105 Steam Cylinder Head.
106 Valve Chest Stuffing box
107 Steam Pipe.
108 Steam Piston.
109 Valve Rod Head.
110 Long Valve Rod Link.
111 Short Valve Rod Link.

112 Hand Hole plate for cap.
113 Water Cap.
114 Water Cylinder Head.
115 Water Plunger Bushing.
116 Steam Valve.
117 Plunger Bushing Ring.
118 Air Chamber and Tee.
119 Piston Rod.
120 Long Crank.
121 Short Crank.
122 Steam Cylinder Foot.
123 Lubricator.
124 Valve Rod.
125 Drip Cock.
126 Hand Hole Plate for W.C.
127 Water Plunger.
128 Plunger Bushing Stud.
129 Piston Rod Gland.
130 Piston Rod Link Arm.
131 Piston Rod Nut.
132 Piston Rod Link.
133 Piston Rod Check Nut.
134 Oil Cup.
135 Valve Rod Gland.
136 Long Lever.
137 Short Lever.
138 Pin for Valve Rod Head.
139 Lever Pin.
140 Link Arm Pin.
148 Water Valve Seat.
149 Water Valve Stem.
150 Water Valve Spring.
151 Water Valve.
152 Water Valve Cover.
162 Steam Piston Head.
163 Steam Piston Follower.
164 Steam Piston Packing Rings.
165 Steam Piston Wedge.
166 Steam Piston Packing Screw.
167 Steam Piston Spring.
168 Valve Chest Cover.
169 Bearing Stand.

Ques. What instructions must one give in ordering a steam pump? Ans. In ordering a pump, the buyer should inform the parties of whom he orders a pump, the following points: 1. For what purpose the pump is used. 2. The nature of the liquid to be pumped, hot, cold, salty, fresh, clear or gritty, also the largest quantity required to be pumped per hour. 3. To what height lifted and what height forced. 4. The ordinary pressure of steam used. P. S. When ordering parts for pump in use, order by proper name, also stating size of pump and manufacture, whether new or old style.

Ques. Could water be forced into the boiler if there were three or four check valves on the discharge pipe? Ans. Yes, water could be forced through all, but it would be more labor on the pump.

Ques. Where is a pet cock put on the pump barrel for cold water, and why? Ans. It is put at the side and near the bottom of the pump barrel, and is there to show how the pump is working, and to drain the pump in winter to prevent it from freezing.

Ques. How is it known when the pump is in good

working order? Ans. By opening the pet cock and noticing the stream that comes out.

Ques. How does the stream show when the pump is in good working order? Ans. Nothing in the suction stroke and full force in the discharge stroke. (Single action.)

Ques. Where would the trouble be if water came full force both strokes? Ans. The trouble would be located at the check and discharge valves, both being caught up.

Ques. Where would the trouble be located if water came full force both strokes, moderate, tank or hydrant pressure? Ans. At the receiving valve.

Ques. Can a pump work without a check valve? Ans. If the discharge valve in the pump is in good order, it can; but if there is neither check nor discharge valve, it can not.

Ques. Can a boiler be fed without a pump? Ans. If the pressure of the boiler is below the pressure of the feed water or city pressure, it can, by simply opening a water valve and letting in the amount of water required.

Ques. By what other means is a boiler fed? Ans. By an injector or an inspirator.

Ques. What is an injector or an inspirator? Ans. They are devices to answer for a pump in feeding a boiler; they draw force and heat the water at the same time.

TO CONNECT THE INSPIRATOR.

In all cases connect so it will take steam from the highest point of boiler. Place a globe valve in steam pipe, just above the inspirator; a globe valve in the supply pipe, close to the Inspirator, and a Check and Globe valve between Inspirator and boiler. If the feed is delivered through a heater, place a check between it and the Inspirator.

Blow out steam pipe before connecting. For a high

lift or long draft, make the suction one size larger than

HANCOCK INSPIRATOR.

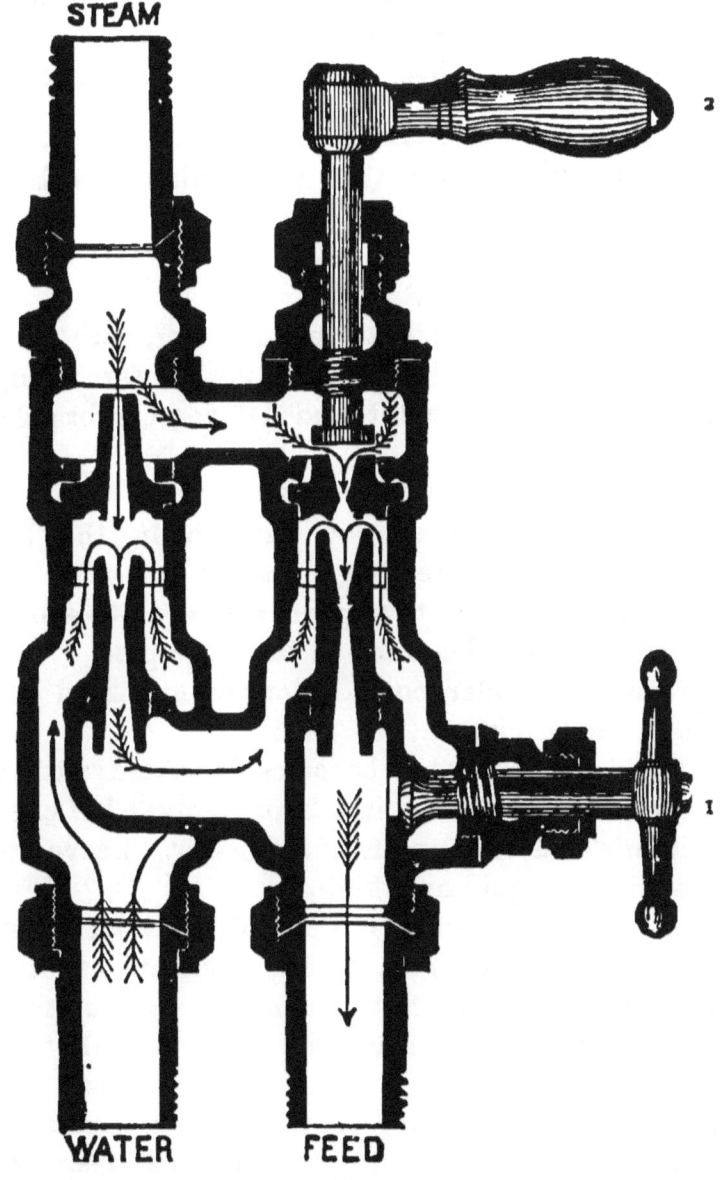

OVERFLOW VALVE UNDER FEED.

TO START THE INSPIRATOR.

See that the overflow valves, marked 1 and 3 are open and the forcer valve marked 2, is closed. Give full steam. After getting water, close No. 1, open No. 2 one quarter of a turn, close No 3, and the Inspirator is at work.

---

the connection. Be sure that the section pipe is absolu ely tight.

The Conditions must be:

First. An Air tight Suction.

Second. An abundant supply of water, with a lift not exceeding 25 feet, and a temperature not exceeding 140 degrees Fahr. for a low lift, and 110 degrees for a 25 foot lift.

Do not connect with other steam pipes.

Tap the boiler where you can obtain the Dryest Steam, and if you are obliged to connect with a large steam pipe, tap it on the upper side so as to avoid the drip caused by condensation in the large pipe.

Do not allow boiler compositions of any kind to pass through the Inspirator.

In case the Inspirator becomes incrusted with lime disconnect it and place it in a bath composed of one part of muriatic acid and 9 parts of soft water. Leave the Inspirator in the liquid over night.

Ques. State the principal upon which a jet of steam taken from the boiler at boiler pressure can force a stream of water back into the boiler through the Injector or Inspirator? Ans. It acts upon the principal of a light body moving at a high velocity giving a slower motion to a heavier body effecting an entrance by means of the momentum thus given to it. For instance, steam at the pressure of 80 lbs to the square inch will escape into the air with a velocity of 1,821 feet per second or 1,241 miles per hour. This rapidly moving jet of steam causes, at first a vacuum in the casing of the injector which fills with water. The steam then mingles with the water, condenses and imparts its velocity to it. The stream of water is then forced along the pipe and strikes the check valve with a force sufficient to open it and then enters the boiler.

Ques. Will the inspector work if the water that is supplied is to hot to condense the steam? Ans. No.

Ques. Why? Ans. Because steam is highly elastic and bulky, and, of itself, would have no effect in driving the hot water in any particular direction. But when steam is moving at a high velocity and is condensed, these particles of water have the power of driving the main body before it into the boiler.

EXCELSIOR BOILER FEEDER.

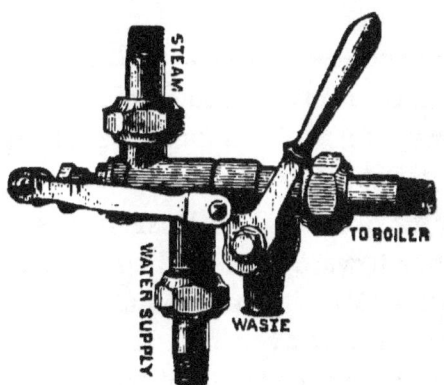

The principal is easily explained, for instance; if a block of wood is laid upon the water it will float, but if it is thrown violently downward it will at first go below the surface. Then if there were something there to catch it and hold it, we would have a state of affairs similar to the injector, where the water enters the boiler by its own momentum and is held there by the check valve.

DIRECTIONS FOR CONNECTING TO THE BOILER.

Take the dry steam from the highest part of the boiler and connect to the coupling on top of the feeder, placing a globe valve on this at any point most convenient for the user to work it. This valve should be kept in GOOD ORDER SO AS NOT TO BE LEAKING STEAM WHEN NOT IN OPERATION. Attach the feed pipe to the coupling on end of feeder taking to boiler opening, place a good check valve at this point near to boiler as practicable. It is desirable in laying suction pipe to commence at the well or tank, make as few joints as possible, AS THE WATER LINE MUST BE AIR-TIGHT, otherwise the feeder will not work; connect this to the coupling on bottom of feeder.

If the lift or draw from water supply is over ten feet distance the pipe should be one size larger than the connections on feeder.

Then remove the feeder from the couplings and blow out steam pipes to remove iron, scales, and lead used in the connections, etc. This should be observed and save the dirt from getting into the jets of feeder.

TO PLACE THE FEEDER IN POSITION FOR WORKING.—Push the lever towards the feed line till it stops.

TO OPERATE.—Give it full steam by the steam valve; when water appears at overflow, slowly reverse position of lever to stop, when feeder is at work.

OPERATE UNDER PRESSURE.—In supplying the Injector from a hydrant or tank pressure, place a globe valve in the water pipe to regulate the quantity of water delivered to the Injector. When the water pressure is very great, this valve should be partly closed, to give satisfactory results.

Ques. Must a pump have valves? Ans. Yes if a

pump did not have any valves it could not do any work. A pump is not a pump unless it has a valve.

Ques. Name the different pumps? Ans. The common well hand pumps with one valve, called a receiving or suction valve, the force or common plunger pump with two valves, a receiving and discharge.

Ques. Explain the working of the valves? Ans. The discharge is to retain the water after it is delivered, so that the plunger can get a fresh supply. After the plunger has ascended and begins to descend, the water sets on top of the receiving and under the discharge; consequently, when the plunger descends it forces the receiving shut and the discharge open.

Ques. Give the proper lift for check valves on feed pipes. Ans. The duty has much to do with lift of valve, for instance, pumps used for pumping under very high pressures, the valves are given a very little lift, thereby lessening the shock and prolonging the life of the valves.

Ques. Should there not be another valve between the check valve and the boiler? Ans. Yes, a globe valve.

Ques. Why is it put there? Ans. To close and keep the pressure in the boiler, in case the check valve is caught up and needs repairing.

Qeus. Can a pump raise, lift or suck hot water? Ans. Not very well.

Ques. Why? Ans. Because the pump would get steam bound. Hot water should be level or higher than the pump in order that the pump should work well.

Ques. Where should a pet cock be put on the pump barrel for hot water? Ans. At the top of barrel, immediately under the packing ring. (Plunger pump.)

Ques. Why is it put there? Ans. To let out steam when steam bound, and air when air bound. There should be a pet cock tapped in the cap of the valve chamber to let off the steam or air when steam or air bound.

Ques. If there was no pet cock on the valve chamber cap, what could be done? Ans. Simply take a wrench and loosen one of the chamber cap nuts a little until the air or steam was out, then tighten it again.

Ques. Why is an air chamber put on a double action pump, and what is it? Ans. It is simply a copper vessel air tight. When the pump is working, the water is forced up into the chamber, which compresses the air, and the compressed air acts as a cushion on the valves and piston head in the water cylinder.

Ques. What is meant by the term cushion? Ans. A cushion is anything that is compressed, and by its compression is formed into a higher and stronger pressure, acting as a spring, deadening any knock that might occur. In a pump, water will cause a knock, it being as solid as iron minus the air, so if a double action pump had no air chamber there would be a continual knock.

Ques. What is known by the term vacuum? Ans. A vacuum is a space void of matter.

Ques. Can a perfect vacuum be created? Ans. No, about 9 to 11 per cent. of the atmospheric pressure which is 14.7 pounds per square inch.

Ques. What will a vacuum do? Ans. It is supposed to lift water 33 feet, providing all pipes and connections are air tight. (The best known lift is 29½ feet.)

Ques. How is a vacuum created or made? Ans. When the plunger of a pump is well packed and it pulls it excludes the air out of the pump barrel and suction pipe, consequently the water, being at the other end of the pipe, it follows the plunger; or, in other words, the atmospheric pressure, being 14.7 pounds per square inch, forces the water up the pipe to fill the vacancy made by the plunger forming the vacuum

Ques. What should be placed a the bottom of the suction pipe? **Ans.** A strainer made out of gauze wire, a foot valve and a pet cock to drain it.

Ques. If the pump was not working and the water running low, and you were asked to run awhile longer would you run and let the water become dangerously low? **Ans.** No, take no chances whatever, but shut down and go about finding the trouble.

Ques. Where would you look for the trouble? **Ans.** Open the pet cock of the pump, and that will nearly tell where it is; if no water came, the water is shut off, or there is none, etc. (as stated before.)

Ques. What prevents a pump from working? **Ans.** Not enough water, too small a suction pipe, the obstruction of the valves to seat by straws, sticks or anything that may be drawn through the suction pipe, or the valves sticking.

Ques. If an accident happened such as a broken pipe connecting the boiler and pump, so that sufficient water could not be had to supply the boiler, what should be done? **Ans.** Simply shut down the engine and all valves connected with the boiler, draw the fire, raise the flue caps, and close the damper.

Ques. Why so. **Ans.** To keep what water there is in the boiler until the trouble is found and repaired.

Ques. If the suction pipe should spring a leak, what should be done? **Ans.** Take a piece of sheet rubber, some copper wire, wrap around tight, and stop the leak temporarily.

Ques. It the hydrant, that supplies the pump with water, should happen to get broken, what should be done? **Ans.** First see how much water there is in the boiler, by trying the gauge cocks, then shut off the water in the street, or wherever the lazy cock lay, and try. if possible to repair it. If an injector or inspirator was attached, and had a tank or well supply, use either until the break is repaired.

Ques. For instance, if there were neither of these, what should be done? Ans. Shut down the engine, close the damper, raise the flue caps and draw the fire, then there would be no danger.

Ques. What is a gravity steam trap. Ans. A device for returning the water of condensation for a heating system to the boiler by gravity.

Ques. Explain its construction and operation? Ans. The trap is usually spherical, containing a float which rises as the water accumulates, at its full heighth. The float closes the return valve and opens boiler connections, both steam and water, equalizing the pressure. The trapped water being a trifle above the water line in the boiler will seek its level and escape into the boiler at its complete discharge, the float falls closing all the boiler connections and opening the return drip and air valve at the top of trap allowing the drip to refill the trap again, etc.

Ques. How high should a valve lift to clear itself? Ans. About one-fourth of its diameter or one-third of its area.

Ques. What proportions should the valves be to any sized pump? Ans. They should be one-fourth the area of the pump itself.

Ques. Suppose in the evening when you shut down that the pump was in good working order, and when you started up the next morning and opened the pump pet cock a strong stream of water came out both strokes, where would you locate the trouble? Ans. The trouble would be at both the check and discharge valves being caught up.

Ques. Suppose you started the pump and it was in good order, and no water came; where would you locate the trouble? Ans. The suction pipe is leaking out of water, or there is no water.

Ques. State the usual area proportion of the cylinders of the steam pump? Ans. The steam cylinder averages four times the area of the water cylinder.

THE ENGINE.

Ques. What is a steam engine? Ans. A steam engine is a machine through which power is obtained from steam.

Ques. What is steam? Ans. Steam is a gaseous vapor evaporated from water, by heat, and is composed of hydrogen and oxygen.

Ques. How do you know water is composed of hydrogen and oxygen? Ans. Science shows that 1 pound of hydrogen with 8 pounds of oxygen is equal to 9 pounds of water.

Ques. What is an engine composed of? Ans. A bed plate, cylinder, connecting rod, crank, crank-shaft, main-pillow block, out pillow block, tail block, cross-head, wrist-pin in cross-head, crank-pin, two cylinder heads, piston-rod, piston-head, follower head, bull-ring, packing rings, follower plate and bolts, connecting rod and brasses, pillow-block brasses, a valve, and guides where the cross-head slides in, so the piston is kept central in the cylinder. The main pillow-block brasses are generally made into four pieces, called top, bottom and two quarter brasses; they are made into four parts, so as to take up the lost motion.

Ques. What precaution should be taken in starting engines? Ans. All engine cylinders should be well drained and heated before starting, then the engine should be started slowly, as the water that accumulates

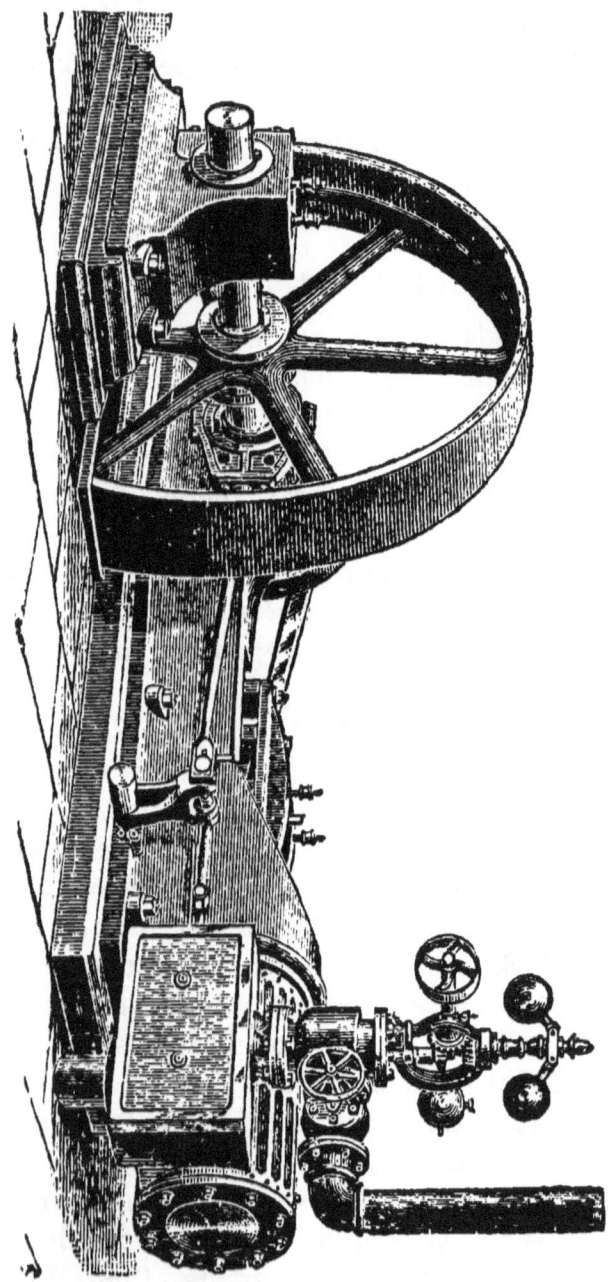

in the cylinder may injure the piston, cylinder, or cylinder heads. Always leave the cylinder cocks open when not running, and they should remain so until the cylinder is heated by the steam,—after the engine has been running at full speed about 5 minutes or such a matter.

Ques. What keeps the rod from running off the crank pin? Ans. The outside shoulder on the crank-pin.

Ques. If water should accumulate in the cylinder, what would be the consequence? Ans. It is liable to crack the cylinder and disable the engine, also cause loss of life.

Ques. If you had charge of an engine in the country, and the cylinder head should happen to crack, how would you remedy it in a hurry? Ans. If not broken too bad, try to patch it with pieces of iron or boards, and brace it from the wall with a piece of heavy scantling, then try and run the engine until a new cylinder head could be made, or make a wooden cylinder head temporarily.

Ques. What size should a steam pipe and an exhaust pipe be to any size cylinder? Ans. The steam pipe should be one-fourth and the exhaust pipe one-third the diameter of the engine cylinder itself.

Ques. If your crank pin or other journals became hot, what would you do? Ans. Try, while running, to get water on them, then oil them; if that would not do, stop and slack up the key a little, then start up again.

Ques. If the cylinder had shoulders inside, or was out of a true circle, what should be done? Ans. Bore it, or have it bored out.

Ques. In case the throttle valve should become loose from the stem and prevent the steam from entering the valve chest, what should be done? Ans. Close the valve next to the boiler, if there is one; if not, let the boiler cool down, then take the valve and stem out and repair it.

Ques. If the slide valve was not steam-tight, what should be done? Ans. Have the valve planed, then chip, file and scrape the seat to a full bearing.

Ques. If the crank and wrist-pins are worn out of true, what should be done? Ans. Caliper and file them until they are round and true.

Ques. What causes the wrist-pin in the cross-head and crank-pin to wear the way they do? Ans. It is simply the motion they have; the crank goes all the way round, forming a circle, and the wrist only vibrates.

Ques. If the cross-head or crank-pin brasses were brass-bound, what should be done? Ans. They should be chipped and filed.

Ques. How do you know when you have taken off enough? Ans. By outside and inside calipers.

Ques. How does steam enter the cylinder? Ans. In common slide-valve engines it enters through one of the end ports and exhausts back through the same port, when the cavity of the valve has covered it and the exhaust port at the same time.

Ques. What is meant by clearance? Ans. Clearance is the space between the piston head, cylinder head and valve face at each end of the stroke.

Ques How would you know the amount of clearance there was in that space? Ans. By finding the number of cubic inches in a bucket of water, then fill up the space level with the steam port, and see how much water is left in the bucket; the difference is the contents in cubic inches.

Ques. Why are gibs, keys and set screws used on both ends of the connecting rod? Ans. They are there to take up lost motion.

Ques. How is it done? Ans. By loosening up the set screw, and driving down the key; then tighten the set screw to keep the key from raising.

The Gardner Governor.

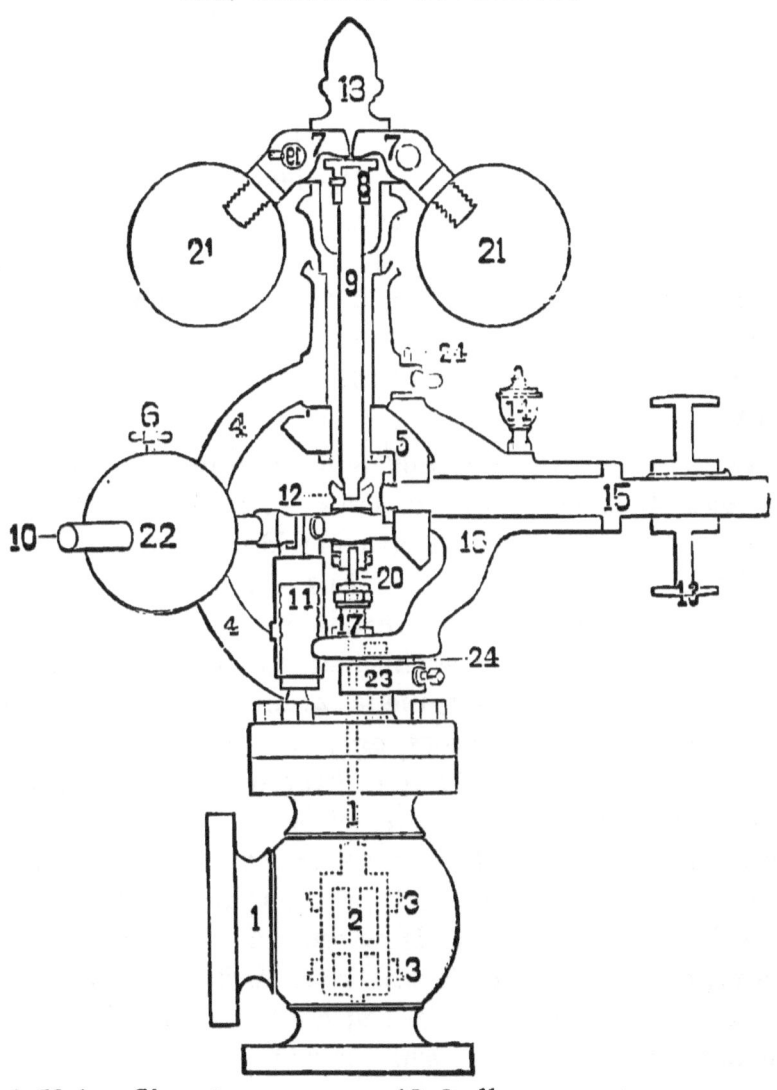

1 Valve Chamber.
2 Valve.
3 Valve Seats (2).
4 Frame.
5 Gears Mit·e.
6 Lever Ball Screw.
13 Pulley.
14 Oil Cup.
15 Pulley Shaft.
16 Shaft Bearing.
17 Stuffing Box.
18 Head.

7 Arms (2).
8 Toe Plate.
9 Spindle.
10 Lever.
11 Fulcrum and Stud.
12 Step Bearing.

19 Arm Pins (2).
20 Valve Stem.
21 Arm Balls.
22 Lever Ball.
23 Check Stud.
24 Sleeve Centers (2).

Ques. Are there more square inches in one end of the cylinder than in the other? Ans. In one sense of the word there are, and in the other there are not, as the piston rod takes up some of the space in one end of the cylinder, therefore there is not the same area in one end as in the other.

Qu s. What is a governor or an engine for? Ans. It is to regulate the steam that passes from the boiler to the steam chest, when the throttle is wide open.

Ques. How does it work? Ans. It is regulated to allow the engine to run at a certain speed. The governor has a belt from the main shaft to a pulley (13) on the governor. After the engine is running up to the speed it is intended to, it allows only enough steam to enter through the governor valve to keep up the same speed; if the engine needs more power it begins to slack up, the governor balls drop, the valve opens and allows more steam to enter; consequently, the engine must retain its speed; and if the load is taken off it will start to run away, the governor balls will rise, and force the valve shut, and cut off the steam; consequently the engine must come back to its regular speed.

Ques. Are there other makes of governors? Ans. Yes, the Automatic Governor.

Ques. What is a lubricator? Ans. A lubricator is an appliance for holding oil, to be d stributed into the valve chest and cylinder, to prevent cutting.

Ques. How is it operated? Ans. It is operated by steam forcing the oil out of the lubricator into the steam pipe.

DIRECTIONS.

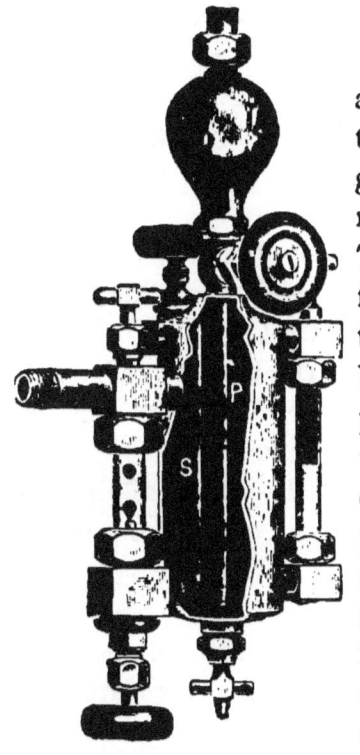

How to apply.—First, drill and tap the steam pipe above the throttle with ½ or ¾ inch gas tap as may be required to receive the oil discharge pipe. Then tap the steam pipe three feet or more above the top of the condensing chamber as may be required, using ¼ inch gas pipe for steam connecting tube, which attach to top of condenser. If, for any reason, the steam pipe can not be tapped three feet or more above the condensing chamber, it may be tapped lower down and the tube of required length may be bent or coiled.

Description of the Sectional cut:
P. is the condensed water pipe.
S. is the oil pipe.

How to fill and operate.—Close valves D and E, open valve G, draw off the water, close valve G and fill with oil. First, open valve D, then regulate the flow of oil with valve E.

In case of strong pulsation, valve N to be partially closed until oil feeds steadily; same to be closed in case of breaking feed glass.

Before starting the cup, time should be allowed for sight-feed glass and condensing chamber **to fill with water by condensation**.

When there is danger from freezing, when not in use valves D, G and E should be left open.

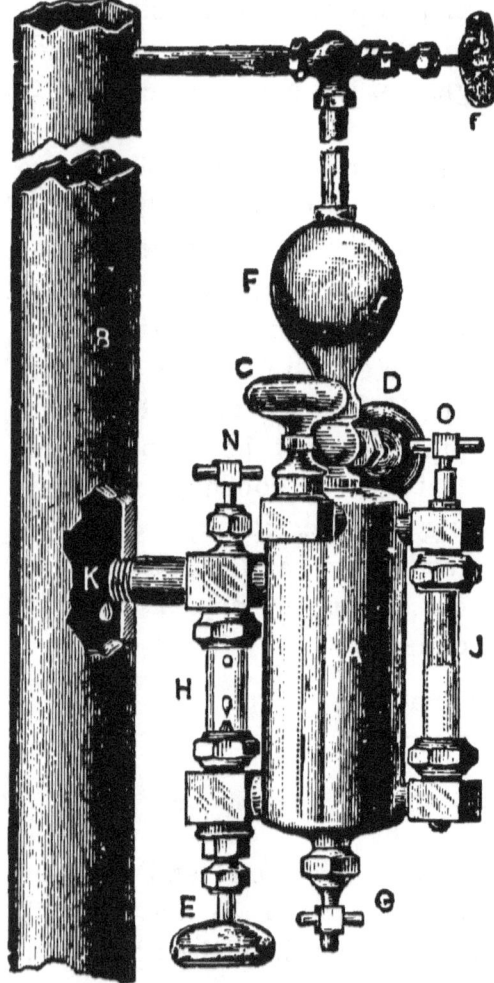

DESCRIPTION.

A—Oil Reservoir.
B—Steam Pipe.
C—Oil Filler.
D—Water Feed Valve.
E—Valve to regulate flow of oil.
FF—Steam Tube and condensing chamber.
G—Drain Valve to draw off water to prevent freezing etc.
H—Sight Feed Glass.
J—Glass Indicator.
K—Oil Discharge pipe.
N—Valve to correct pulsation or unsteadiness in feed.
O—Vent.

POUNDING ENGINES.

SOME OF THE CAUSES THAT LEAD TO THIS TROUBLE.

Since engines of high speed and short stroke have come into general use, it is not uncommon to find many of them quite noisy. The engineer in charge of a smoothly running engine takes great pride in showing it, but when a click or pounding is heard, he naturally feels that these defects reflect upon his skill, and whether alone or in the presence of visitors, it is a constant source of annoyance. Many hours of overwork for which he makes no charge, are employed in trying to locate the cause of these noises, which to outsiders are of no consequence, but to his ear become almost unbearable. Sometimes, when the main belt is off, the engine runs so smoothly he fancies the trouble has been reached, but in the morning when the machinery is set to work the exasperating pounding begins, and to aggravate the case, boxes that formerly had given no trouble, begin to heat from being too closely adjusted.

There are few, perhaps, outside of the engineering profession who know how perplexing a thud, or pounding in an engine becomes. Go to the cylinder and it seems to be there; stand at the crank and the noise is there. The sensitive ear of the engnieer can hear nothing else, and its continuance affects both mind and body in a way that is hard to explain.

It would be as easy to prescribe one remedy for all cases of dyspepsia, as to give a rule for finding the causes of pounding in an engine. It is well known that a want of

PROPER ALIGNMENT

is one of the most common causes, and every engineer worthy of the name should be able to adjust the working parts of an engine to a line. Aside from imperfect workmanship in the erection of an engine, there are other causes of pounding to which we may call attention, the obscurity of some of these, may cause them to be overlooked. Wrist-pins in both cranks and cross-heads naturally wear unevenly, and these defects are frequently allowed to go on until the boxes will become quite loose at the dead centers, and tight at the quarter centers. Calipers applied to the wrists will detect these defects. Another defect in wrist-pins, not so easily detected, is faulty workmanship, the wris's not being placed squarely in the crank or cross head. A good spirit level will detect the slightest deviation in this and may be applied as follows: Disconnect the rod from the cross head and tighten it to the crank pin so it may turn, but not vibrate sidewise; place the rod in a position to move freely as the crank is turned; attach a spirit level to the rod with a clamp and in line with the main shaft. It matters not whether the main shaft be leveled. The position of the bulb in level should not change when the crank is revolved. But if the wrist pin is not set squarely, the level will be tipped from side to side, as the crank revolves, and the places the bulb occupies at different points in the revolution will indicate the direction the wrist takes from that of a correct position. It will be readily seen that this test detects the slightest discrepancy, as the deviation is doubled by this reversing of the position of the wrist in its revolution. It may be a matter of surprise in thus examining wrist-pins, to find how many there are that are not set

PERFECTLY SQUARE.

The power of such a wrist to strain, and open its boxes is almost unlimited and most seriously affect the running of an engine.

Another fruitful cause of unsteady motion and noise in the running of an engine, is improperly balanced cranks. Much has been written on this subject, so we will leave it by merely saying that a disk crank, in order to hold the proper amount of balance, should be several inches larger than is needed for the stroke of the engine. This we learned a number of years ago, and have since seen that many builders of high speed engines have adopted this plan.

Pounding is sometimes aggravated, or in fact produced, by improperly proportioned valves. Especially is this so when engines have high speed and heavy reciprocating parts. The questions involved are of vital importance in the construction of a good and economical engine, but they are too intricate to be treated here. The one feature to which we call attention is that of the cushion. This is to confine enough exhaust steam in the cylinder at the termination of each stroke to form a spring as it were, to receive the impact of the piston as it is brought to a stop. If this force is not spent on such a spring, it evidently falls on the boxes in the rod and main shaft. If such faults as these exist, they may never come to notice, but if they should, the engineer has no remedy, unless it be by patching the old valves, or introducing a new one. So he is not always to be censured when his engine does not run smoothly, for these defects may indeed be chronic and have their origin in the designing room or in the workshop.

Ques. What is the best way to get a live shaft in line without taking the shaft out? Ans. The best way is to use specially prepared instruments, but in absence of these a very good way is to run a line along, say 8 inches to one side of the shaft and paralled to its centre; then by the use of an ordinary level bring the shaft up level, and for side alignment use a piece of wood with one end formed to fit the shaft and the other brought to a point. By placing the curved end against the shaft and bringing the shaft so that the pointed end of the stick will just touch the line along its entire length will secure a nicely running shaft provided that the couplings are bored out in line. Very often a poorly lifted shaft in the coupler is the cause of a badly running shaft, requiring extra power to drive it and no amount of lining up will remedy the evil.

LINING AN ENGINE.

Ques. How would you line up an engine? Ans. By stripping the engine, take off both cylinder heads, if convenient; then take out the follower head, piston rings, bull-ring; disconnect the piston from the cross-head; also disconnect the connecting rod from the cross-head and the crank pin; then take a slotted stick and place it on one of the studs on the end of cylinder furthest from the crank, then draw a fine sea-grass line over the stick and through the centre of cylinder, and attach it to an upright stick at the other end of the bed-plate, nailed to the floor or clamped to the bed-plate; then take a thin stick, (like a lead pencil) the length of it being a half inch less than half the diameter of cylinder, and stick a pin in each end of the stick, so they can be forced in or drawn out to suit the adjustment; then centre the line at each end of the cylinder at the counter-bore from four sides. Never centre the line in the stuffing box where the piston passes through, but use the inside counter bore under all circumstances, whether you can remove the back cylinder head or not. Some engine cylinder heads and frames are one; consequently, the head can not and must not be moved.

Ques. If one counter-bore was larger than the other what should be done? Ans. Centre it accordingly, by using two centering sticks.

Ques. Why is the counter-bore used? Ans. Because the counter-bore is the only true bore the cylinder has that is not worn; consequently, all engineers and machinists must be governed by it in lining.

Ques. What is a counter-bore? Ans. A counter-bore is each end of the cylinder bored from one sixteenth to one fourth of an inch larger, from 1 to 5 inches long, according to the size and length of the cylinder.

Ques. Of what use is a counter-bore? Ans. To keep the piston from wearing a shoulder in the cylinder at each end.

Ques. Why is it that the counter-bore prevents the piston from wearing a shoulder in the cylinder? Ans. Because the piston rings just pass over the edge of the regular bore, and by so doing no shoulder can be formed in the cylinder.

Ques. How are cylinders bored? Ans. They are generally bored on a regular cylinder boring lathe, which has a table that can be raised or lowered to suit. The regular bore is first bored, then the counter-bore, then the two faces for the heads.

Ques. How is a shaft squared when the line is centrally through the cylinder? Ans. By moving the crank-pin down to the line and seeing where the line touches the crank-pin between the two shoulders, then move the pin over to the other dead center, and see how it comes; if equal, the shaft is square.

Ques. If it was out of square ¼ of an inch, what should be done? Ans. Move the out end pillow-block (or tail block.)

Ques. Why not move the head-block? Ans. Because it would alter the length of the connecting-rod, and be liable to knock out a cylinder-head.

Ques. How would you level a shaft? Ans. A shaft is leveled by a spirit level, or a plumb-line dropped past close to the line that comes through the cylinder directly in front of the center of shaft; let it drop in a bucket of water to keep the plumb-bob from swaying around; then try the crank pin at both half strokes (the same principle as in squaring), top and bottom, and see how the crank pin feels the line ; if equal, the shaft is level.

Ques. If the shaft was out of level, what should be done? Ans. Simply thin or thicken the brasses, or babbitt the main pillow and out block bearings, whichever the case may be.

Ques. How is it known if the center of the shaft is in line with the line through the cylinder ? Ans. it can be found out by placing a two-foot steel square against the crank face, under the line through the cylinder, so that the heel of the square is at the center of the shaft, and see how the square touches the line ; if it touches exactly, the shaft is in line; if too hard, the shaft is too high ; if not at all, the shaft is too low.

Ques. How is the shaft raised? Ans. There are various ways: by liners, babbitt, heavier or lighter brasses.

Ques. If the crank was oval and a square was put against it, would that be right? Ans. A spirit level could be placed on the square and bring it level, or drop a plumb-line, and put the end of the square against the crank-shaft center, and let it come against the plumb-line. This is a very true way.

Ques. Now, after the shaft is in line, square and level, and is still out over the line ¼ of an inch, how could it be remedied? Ans. Simply take it off the crank pin brasses and fill in the other side with a brass ring, or babbitt the side edge of brasses; in some cases the sides of the connecting rod has to be chipped to allow it to pass free of the crank-face.

Ques. Why could it not be taken off the wrist-pin brasses in the cross head? Ans. Because the rod would then be out of the center of the cross-head and have a tendency to bind the piston in the cylinder and the cross-head in the guides, consequently cutting both.

Ques. Would it not make a difference at the other end of the rod? Ans. No, the closer the crank face the better it would be.

Ques. Now what should be done? Ans. Level and line the guides by putting them in their place, and line them with a pair of calipers, by calipering them at both ends to get them in line with the line through the cylinder, after having found the distance between the side of the cross-head and the center of the cross-head where the piston enters the cross-head. Level by spirit level, first taking spirit level and trying it in cylinder, if a new one, or on top of the cylinder where it has been planed off when first bored, for they are the only things to go by.

Ques. Could the valve seat be used to level by? Ans. No, but alongside of it, where the steam chest rests on.

Ques. If you had no spirit level, how would you do it? Ans. With a plumb-line, by placing a square lengthwise on the guides, and try them by bringing the square against the line.

Ques. If no two-foot square could be had, how could one be layed off Ans. Take a pair of dividers, draw a circle, then find four points on the circle, scribe lines from point to point, which gives a square. This should be done very accurately, or 6,—8 and 10 a triangle.

Ques. Explain the use and figures on a steel square? Ans. The standard steel square has a blade 24 inches long and 2 inches wide, and a tongue from 14 to 18 inches long and 1½ inches wide. The blade is exactly at right angles with the tongue, and the angle formed by them an exact right angle, or square corner. A proper square should have the ordinary divisions of inches, half inches, quarters and eighths, and often sixteenths and thirty-seconds. Another portion of the square is divided into twelfths of an inch; this portion is simply a scale of 12 feet to an inch, used for any purpose, as measuring scale, drawing, etc. The diagonal scale on the tongue near the blade, often found on squares, is thus termed from its diagonal lines. However, the proper term is *centesimal scale*, for the reason that by it a unit may be divided into 100 equal parts, and therefore any number to the 100th part of a unit may be expressed. In this scale A B is one inch; then if it be required to take 73-100 inches, set one foot of the compass in the third parallel under one at E, extend the other foot to the seventh diagonal in that parallel at G, and the distance between E G is that required, for E F is one inch and F G 72 parts of an inch.

Upon one side of the blade of the square, running parallel with the length, will be found nine lines, divided at intervals of one inch into sections or spaces by cross lines. This is the plank, board and scantling measure. On each side of the cross lines referred to are figures, sometimes on one side of the cross line and often spread over the line thus: 1 | 4—9 | . We will suppose we have a board 12 feet long and 6 inches wide. Looking on the outer edge of the blade we will find 12; between the fifth and sixth lines, under 12, will be found 12 again; this is the length of the board. Now follow the space along towards the tongue till we come to the cross line under 6 (on the edge of the blade), this being the width of the board; In this space will be found the figure 6 again, which is the answer in board measure, viz., six feet.

On some squares will be found on one side of the blade 9 lines, and crossing these lines diagonally to the right are rows of figures, as seven 1s, seven 2s, seven 3s, etc. This is another style of board measure and gives the feet in a board according to its length and width.

In the center of the tongue will generally be found two parallel lines, half an inch apart, with figures between them; this is termed the Brace Rule. Near the extreme end of the tongue will be found 24-24 and to the right of these 33-95. The 24-24 indicate the two sides of a right-angle-triangle, while the length of the brace is indicated by 33-95. This will explain the use of any of the figures in the brace rule. On the opposite side of the tongue from the brace rule will generally be found the octagon scale, situated between two central paralell lines. This space is divided into intervals and numbered thus: 10, 20, 30, 40, 50, 60. Suppose it becomes necessary to describe an octagon ten inches square; draw a square ten inches each way and bisect the square with a horizontal and perpendicular center line. To find the length of the octagon line, place one point of the compasses on any of the main divisions or the scale and the other leg or point on the tenth subdivision. This length being measured off on each side of center lines, touching the line of the octagon will give the points from which to draw the octagonal lines. The size of the octagon must equal the number of spaces taken off from the tongue by the compasses.

Ques. Can a plumb-line hang out of true? Ans. It can not, provided it hangs clear of everything. If none of these were handy, a straight edge must be placed across the guides at one end, and see if the guides touch the straight edge equally at both edges, then caliper the distance between the line and the straight edge, also at the other end of the guides; if the same, the guides are level lengthwise with the cylinder and line; then level the guides crosswise with a plumb-line and square.

Ques. How is the measure of the connecting rod of an engine found? Ans. By finding the striking points.

Ques. How is that done? Ans. By shoving the piston and cross-head up against the cylinder-head, and making a mark on the guides at one end of the cross-head with a scriber and center-punch; then move the piston and cross-head back to the other cylinder-head and make another mark on the guide at the same end of the cross-head; then measure from the center of the crank-pin to center of the shaft: that gives the half-stroke; double this, gives the full stroke. If half-stroke is 12 inches, the full-stroke is 54 inches; then if the distance between the two striking points is 25 inches, add the stroke 24 inches, the clearance between the cylinder-head and piston-head will be ½ inch when the piston is at either end of the cylinder. Then move the cross-head ½ inch back from the striking point, and bring the crank-pin toward the same dead center; then take a tram and measure from the outside center of crank-pin to the outside center of wrist pin in cross-head, which will give the proper length of the connecting-rod, also the right division of clearance.

Ques. What is meant by the clearance in cylinders? Ans. It is the unoccupied space between the piston-head cylinder-head and valve-face, when the crank-pin is at either dead center.

Ques. Does the amount of clearance affect the engine's economy? Ans. Yes, it does.

Ques. How much clearance should there be between the piston and cylinder-head? Ans. It depends upon the size; Some have from $\frac{1}{4}$ to $\frac{7}{8}$ of an inch.

Ques. What is formed in that space or clearance when running? Ans. A cushion.

Ques. What is a cushion? Ans. A cushion means the steam that enters the cylinder through the lead the valve has, and the resistance it makes on the piston-head cylinder-head and valve-face as the engine is reaching the dead-center.

Ques. What is the cushion for? Ans. It is to catch the piston and weight of the machinery as it reaches the dead-center.

Ques. How is the connecting-rod shortened or lengthened? Ans. By placing tin or sheet iron liners between the brasses and stud-ends of the connecting-rod.

Ques. Now, if the key had to be raised, how could this be done? Ans. By putting liners between the straps and brasses.

Ques. Would that not altar the length of the rod? Ans. No.

Ques. With what tool is a connecting-rod measured? Ans. It is called a "tram."

Ques. With what is an engine packed in the stuffing-box? Ans. Some engineers use hemp, others use black lead packing, and others use lead rings or metallic packing; there are several kinds. Every engineer to his own taste.

VALVE MOTION.

Ques. What is an eccentric? Ans. An eccentric is a subterfuge for a crank; it is anything out of center.

Ques. How is the throw or stroke of an eccentric determined? Ans. By measuring the heavy and the light side; the difference between the two is the stroke or throw.

Ques. What throw should a common slide valve engine eccentric have? Ans. Generally double the width of the entry or steam ports.

Ques. What is a cam? Ans. A cam has no definite meaning; it has 1, 2, 3 or 4 motions; they are used on poppet valve engines, used on high pressure river steamboats.

Ques. How are the valves and eccentric rods of an engine found? Ans. By placing the crank pin at its dead-center, the center of the eccentric straight or plumb above the center of the shaft, the rocker-arm perpendicular, and the valve covering both parts equally; then take a tram and measure from the center of the eccentric to the center of the pin where the eccentric rod hooks on (generally the lower pin) for the eccentric rod, and from the outside center of the pin where the valve rod is attached to the furthermost end of the valve allowing for two nuts at each end of the valve, called adjusting and jamb nuts.

Ques. How is an eccentric brought plumb? Ans. By dropping two plumb lines, one at each side of the shaft, and half the space between the two lines will be where the center of the eccentric should stand, with heavy side up.

Ques. What kind of a tool is used to find the exact center? Ans. A pair of hermaphrodite calipers, one leg of which has a sharp point and the other leg has a short foot, so as to feel the line.

Ques. What does an eccentric rod consist of? Ans. An eccentric rod consists of a strap, yoke, rod and two nuts; when taking the measure, couple the yoke and the strap together, then put a half-inch thick piece of wood between the two straps and find the center of the circle from four sides, with a pair of hermaphrodite calipers, then put the rod in the yoke and adjust it to the proper length by the two nuts; if that will not do, the rod must be shortened or lengthened, by cutting out or adding a piece, whichever the case may be. Then take the measure with a tram from the center of the straps to the center of the rod where the rod hooks on lower rocker-arm pin.

Ques. How long is the thread on a valve-rod? Ans. Long enough to allow two nuts at each end of the valve, and space for adjustment.

Ques. Now, if the rocker-arm stood at a quarter, and the eccentric out of plumb, how could the measure for the rods be taken? Ans. Simply bring them plumb and take the measure; that is the only right way.

Ques. After measuring the rods what should be done? Ans. They should be put on and the valve set.

Ques. How is a valve set after the connections are made? Ans. Move eccentric in the direction the engine is to run, until the valve begins to take steam or lead, then tighten the eccentric temporarily with the set screws, then move the crank-pin over to the other dead center, and see how much lead the valve has; if equal, the valve is set.

Ques. What is meant by the lead of the valve? Ans. The opening the valve has when the piston is at the beginning of its stroke.

Ques. What lead should a large engine have? Ans. About 1-16 of an inch. High speed engines must have a quick opening or good lead.

Ques. Now if you find the valve laps out ⅜ of an inch on one end, and the proper lead on the other, what would you do? Ans. Divide the difference, by moving the valve one-half it is out, by adjusting the valve-gear.

Ques. How much? Ans. The valve has 1-16 of an inch lead at one end and laps ⅜ of an inch at the other end; the valve is out 7-16 of an inch; then the valve must be adjusted by the nuts one-half it is out, making 7-32 of an inch. Then throw the crank on the other dead center, move the eccentric whichever way will bring the valve back to 1-16 of an inch lead, then tighten temporarily with the set screws, throw the crank over on the other dead center, and the valve will be set. After the valve is set, tighten the eccentric for good.

Ques. But if it is not set, what would you do? Ans. Go through the same performance until it is set. Some valve-rods have a yoke that slips over the valve, while the adjusting and jam-nuts are between the stuffing box and the rocker-arm pin. When a valve-rod has no nuts, the adjusting must be done at the eccentric-rod.

Ques. How is the stroke of the valve-rod shortened or lengthened? Ans. To lengthen or shorten the stroke of the valve-rod, raise or lower the eccentric-rod pin in the slot, at the bottom of the rocker-arm, whichever way suits the circumstances.

Ques. Supposing the valve had to be set in a hurry, how could it be done without seeing the valve? Ans. **Simply set the valve by the cylinder cocks.**

Ques. Now after having set the valve, keyed everything up properly, and there was a thud or dead sound in the engine or cylinder, where would the trouble most likely be? Ans. In the exhaust being choked. The steam chest cover should be taken off, then uncouple the valve, turn the valve up sideways and move it until the steam edge has the proper lead with the steam-port, then place a square on the valve-seat of the cylinder, and against the valve-face, to see how the exhaust head on the opposite steam-port corresponds. If it is choked, then scribe it by allowing a little over double the steam lead.

Ques. How is the exhaust made larger in common slide valve engines? Ans. By chipping out the exhaust cavity in the valve, and rubbing a file over it to smooth it.

Ques. Is a little over double the steam-lead sufficient for the exhaust? Ans. Yes; if not, take out a little more.

Ques. Where should the exhaust be? Ans. It should be the furthest from the steam port that is receiving.

Ques. What amount of lap should the valve have to cut off steam at a given point of the stroke of the engine? Ans. Suppose the valve is to cut off steam at $\frac{3}{8}$ of the stroke of the engine; the piston should be moved $\frac{3}{8}$ of its stroke, then see how much opening the valve has; as much opening as the valve has, is the amount of lap required to be added to the valve in order to cut off steam at $\frac{3}{8}$ of engine stroke. This rule answers for all cut offs.

Ques. What should be done in case the eccentric slipped around on the shaft? Ans. Set the valve the same as before.

Ques. Is the principle of valve setting the same on all engines? Ans. Yes, the principle is the same.

Ques. How is the dead center of an engine found?

Ans. By placing a spirit level on the strap that goes around the brasses that connect the crank-pin to the connecting-rod, and when it is level the crank is at a dead center. If the engine is not level, then use an adjustable level.

Ques. By what other way is the dead center of an engine found? By moving the engine toward the dead center until the cross-head stopped moving; then put a center punch mark in the floor, and one on the fly-wheel, after having marked it with a tram; then move the crank over the center until the cross-head begins to move, then put another mark; the middle between the two marks is the exact dead center; then bring the middle mark to the point of the tram; this is done with a small tram with one straight point and a short foot.

Ques. If the engine had to be run in the opposite direction to which it had been running, how could it be done? Ans. It could be done by placing the crank-pin on the dead center, removing the steam-chest cover, and turning the eccentric over on the shaft in the opposite direction, until the valve has the proper lead at the opposite port, then try the engine from dead center to dead center, to equalize the lead at both ends of the valve; then the engine will run in the opposite direction.

Ques. Does a crank-pin and piston travel the same distance? Ans. No, a crank-pin travels 1.1416 times further than the piston each revolution, or 0.5707 times further each stroke. For example, take an engine with a 12-inch stroke, the piston travels 24 inches and the crank pin 37.6992 inches each revolution, or the piston travels 12 inches each stoke and the crank-pin 18.8496 per single stroke of piston. To do this, multiply the single stroke by one-half of 3.1416, which is 1,5708, and the answer will be the distance the crank-pin travels further than the piston per single stroke. This rule answers for all engines. Another fact, not generally known by many men is that a crank of an engine, at two certain points travels a long distance while the motion of the cross-head is hardly noticed. When the center of the crank-shaft and crank-pin are in a line with the piston-rod, no steam pressure applied to either side of the piston can set the engine in motion; this is called the dead center.

Ques. Is the piston-head in the center of the cylinder when the centers of the crank-pin and crank-shaft are plumb, or in right angles with the cylinder? Ans. No, under no circumstances.

Ques. What is a revolution? Ans. It means that the crank has turned once around, or made a circle.

Ques. How many strokes has a revolution? Ans. Two to each revolution.

Ques. If an engine has 36 inches stroke, and makes 80 revolutions per minute, how many feet does it travel in a minute? Ans. 36 inches multiplied by 2 equals 72 inches, this multiplied by 80 revolutions equals 5760 inches, which divided by 12 equals 480 feet per minute.

Ques. If asked the horse power of any sized engine, could you tell it? Ans. Yes.

Ques. Well, how would you go about it, and what is a horse power? Ans. A horse power is 33,000 pounds raised 1 foot high in 1 minute, or 150 pounds raised 220 feet high in one minute. To find the horse power of any engine, first find the area of the piston-head face, then multiply the answer by the average pounds pressure per square inch in cylinder, then multiply by the number of feet traveled in 1 minute, and divide by 33,000.

EXAMPLE:

Cylinder 12 x 24 in.
65 revolutions.

Average pressure 40 lbs.

NOTE.—The *mean* or *average* pressure in the cylinder is *less* than the pressure in the boiler, since the entrance of steam to cylinder s cut off before the stroke is completed. Hence the steam in the cylinder will expand and consequently diminish in pressure towards the end of each stroke. Generally allow about ½ the boiler pressure in figuring the H. P.

```
     12        diam. of cylinder.
     12
     ──
    144        sq. of diameter.
   .7854
   ─────
 113.0976      area op p. h. face.
     40        averag  pressure in
   ─────       the cylinder.
 4523.9040
    260        No. ft. trav. by p.
   ─────
```

33000)1176215.0400(35.9428 I. H. P.

A HORSE'S POWER.

MECHANICAL INTERPRETATION OF THIS UNIT OF MEASURE.

The question is often asked: What constitutes a horse-power It is generally known to be a unit of measure as applied to steam, water, electricity, or any other energy that can be converted into useful effect. Yet the means employed for giving definiteness to the expression are not so generally understood. The term, doubtless, came into use with the introduction of the steam engine, and a mechanical equivalent has been universally accepted, which is expressed in foot-pounds, that is to say, 33,000 lbs. raised one foot in a minute, constitute one horse-power.

To those not familiar with the principles of mechanics this may not be intelligible. Some simple examples and illustrations may, therefore, be allowable. Force or weight, and motion are inseparable in estimating the power which a moving body may exert, and these two factors, being equal in value, their order may be reversed without affecting the result. For example, one pound raised 33,000 feet in a minute represents a horse power as perfectly as if the feet were pounds, as in the first rule given, the power being better suited to the calculation of the power of steam, on account of its great force.

We will first compare the rule with the actual powers of a horse, and then apply it to the steam engine, for which it has been established. The usual traveling gait of a horse, hitched to a light vehicle, is about five miles an hour, or 440 feet per minute. Now, if we attach a scale to the singletree we may note the amount of power the horse is exerting. Assuming this to be 75 lbs. the product of the speed per minute, 440 by 75, gives us 33,000, which in dynamics, is called foot-pounds, and represents a horse-power.

In applying this to a steam engine we have first to determine the area of the piston, that is, the number of square inches it contains; and next, the average pressure of steam applied to it, which is termed the mean effective pressure. The steam gauge at boiler gives no evidence of what this may be but an instrument called an indicator attached to a cylinder, gives it exactly, and it is often a matter of surprise to find only 20 or 30 lbs. per square inch, when the steam in the boiler is held at 80 and 90 lbs. Practice with this instrument has also shown that when ordinary slide valve engines are worked to their maximum capacities, the average pressure on the piston is only about half that of the boiler pressure, and if by reason of a short valve, the average pressure is increased, it is done at a great sacrifice in the

ECONOMY OF STEAM.

We speak of this curtailment of pressure because so many in applying the rules we are about to give, assume too high a pressure in estimating the power of their engines. The next step in the calculation is to determine the speed of the piston when the engine is at its regular work. We will take an engine of 20″ stroke, making 150 revolutions per minute. The piston travels 40″ or 3⅓ feet at each turn. The piston speed is therefore 500 feet per minute. We now have all data necessary for the calculation, except the average piston pressure, which we will assume to be 30 lbs. on a 10 inch piston. The area of this is 78.54 square inches. The formula, then will stand thus.

$$\frac{30 \times 78.54 \times 500}{33,000} = 35.7 \text{ H. P.}$$

It will be seen that the total pressure on the piston is 23.56.2 lbs., which, moving at the rate of 500 feet per minute, make 1.178.100 foot-pounds, and every 33.000 of these is a horse power Hence, to divide by 33,000 gives us 35.7 H. P.

It is sometimes found convenient to omit the first factor of pressure and make this a unit. We then get the H. P. of the engine for one pound to the inch, which in this case is 1.19 horse power. We may now multiply this by any number of pounds we may secure. In case it be 30 we have the result 35.7 as before; if 40 lbs. we have 46 6 horse-power?

THE INDICATOR.

(Thompson's

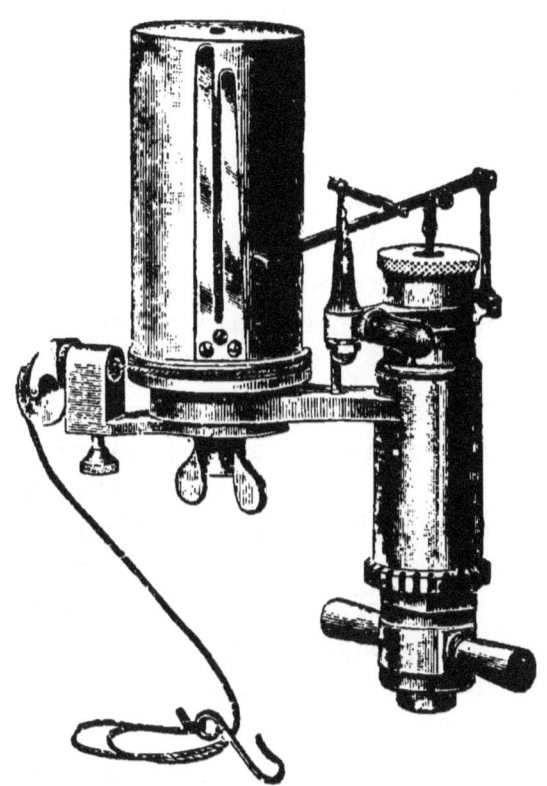

The steam engine indicator is an instrument for showing the pressure of steam in the cylinder at all points of the stroke, or for producing actual diagrams. The indicator consists of a small cylinder accurately bored out, and fitted with a piston, capable of working in the (indicator) cylinder with little or no friction, and yet be practically steam-tight. The piston has an area of just ½ of a square inch, and its motion in the cylinder is 25-32 of an inch.

The piston-rod is connected to a pair of light levers, so linked together that a pencil carried at the center of the link moves in nearly a straight line through a maximum distance of $3\frac{1}{8}$ inches. A spiral spring placed in the cylinder above the piston, and of a strength proportioned to the steam pressure, resists the motion of the piston; and the elasticity of this spring is such that each pound of pressure on the piston causes the pencil to move a certain fractional part of an inch. The pencil in this case is made of a piece of pointed brass wire, which retains its sharpness for a considerable time, and yet makes a well-defined line upon the prepared paper generally used with the indicator.

The paper is wound around the drum, which has a diameter of 2 inches, and is capable of a semi-rotary motion upon its axis to such an extent that the extreme length of diagram may be $5\frac{1}{4}$ inches. Motion is given to the drum in one direction, during the forward stroke of the engine, by means of a cord connected indirectly to the cross-head of the engine, and the drum is brought back again during the return stroke of the engine by the action of a coiled spring at its base.

The conical stem of the instrument permits it to be turned around and fixed in any desired position, and the guide-pulleys attached to the instrument under the paper drum may also be moved around so as to bring the cord upon the drum-pulley from any convenient direction.

The upper side of the piston is open to the atmosphere; the lower side may, by means of a stop-cock, be put into communication either with the atmosphere or with the engine cylinder.

When both sides of the piston are pressed upon by the atmosphere, the pencil, on being brought into contact with the moving paper, describes the atmospheric line. When the lower side of the piston is in communication with the engine cylinder, the position of the pencil is determined by the pressure of the steam existing in the cylinder; and on the pencil being pressed against the paper during a complete double stroke of the engine, the entire indicator diagram is described.

In order that the diagram shall be correct, the motion of the drum and paper shall coincide exactly with that of the engine piston; second that the position of the pencil shall precisely indicate the pressure of steam in the cylinder; third, that the pendulum must be from 1½ to 3 times as long as the stroke of the engine piston; fourth, that the pendulum must be plumb when the piston is at half-stroke; fifth, that the cord around the drum must be attached to the pendulum at right angles, or square with the indicator; sixth, the pendulum must be attached with an inch wooden pin to the ceiling or floor at one end, the other end to the cross-head by means of a screw bolt in wrist-pin and a slot in the pendulum; seventh, that the two holes tapped in the cylinder are directly opposite the steam ports, and centrally between the piston head and cylinder head, when the engine is at the dead center, or, in other words, in the center of clearance; eighth, that the piping should be as short as possible, and ½ inch pipe if not over 1 foot long. If longer the pipe should be larger close to the cylinder, and covered so as not to allow too much condensation, as it affects the diagram. The best way to take a diagram is to tap a hole in each cylinder-head and take each end separately. The cord must be attached to the pendulum, so the paper drum will move in proportion to the piston.

An indicator shows the highest and the lowest pressure reached, also the cut-off and lead. If there is a great difference, say more than 5 pounds, between the boiler pressure and the initial pressure upon the piston, the connecting pipes may be taken as being too small, too abrupt or the steam ports too contracted. The full pressure of steam should come upon the piston at the very beginning of its stroke. Should the admission corner be rounded, the valve is wanting in "lead," or, in other words, the port for the admission of steam is uncovered too late in the stroke.

The steam line shoud be parrallel or straight with the atmospheric line up to the point of cut-off, or nearly so.   (the steam line) fall as the piston advances, the opening tor the admission of steam is insufficient, and the steam is "wire-drawn."

The point of cut-off should be sharp and well defined; should it be otherwise, the valve does not close quick enough. The bevel line leading from the cut-off line to the end of the stroke is called the expansion line.

Q. Which is the standard indicator?

A. The Thompson's improved.

Q. Are there any other makes? A. Yes: Richard's, McNought's, Tabor's and others.

RULES.

RULE for telling the power of a diagram: Set down the length of the spaces formed by the vertical lines from the base in measurements of a scale accompanying the indicator, and on which a tenth of an inch usually represents a pound of pressure; add up the total length of all the spaces, which will give the main length, or the main pressure upon the piston in pounds per square inch; to do this, lay a card taken by the indicator off in ten parts, by drawing lines from top to bottom. Find out what the scale is; suppose it is 60, the number of ordinates 10, and that the sum of their length is 6 inches; so 6 and 10 ordinates = 6-10 or 6 x 60 = 36.0. Ans. 36 pounds pressure upon the piston.

RULE for finding and deducting friction: Multiply N. H. P. by .13 and subtract the answer from N. H. P., which gives I. H. P.

Ques. What is N. H. P? Ans. It is nominal horsepower.

Ques. What is I. H. P? Ans. It is indicated horsepower.

Ques. What is meant by cutting off steam at 6 inches? Ans. It means that the valve closes and cuts off the live steam from the boiler at 6 inches of the piston's travel; then the engine gets its power from the time the valve closes or cuts off until the exhaust opens by the expansion of the steam closed up in the cylinder.

Standard multiplers, with examples:

1. For the Area of a Circle. Multiply sq. of diam. by .7854
2. For Circumference of a Circle, Multiply diameter by 3.1416
3. For Diameter of a Circle, Multiply the circum. by .31831
4. For the Surface of a Ball, Multiply sq. of diam. by 3.1416
5. For the Cubic Inches in a Ball, Multiply cube of dia. by .5236

1. RULE for finding the area of any circle. Always multiply the diameter by itself, then by .7854, then cut off 4 decimals from the right.

2. RULE for finding the circumference of anything round. Multiply the diameter by 3.1416, and cut 4 decimals.

3. RULE to find diameter of circle. Multiply circumference by .31831.

EXAMPLE: The circumference 9.4248 x .31831 = 3.000008088 = 3 inches diameter.

4. RULE to find the surface of a sphere, globe or ball.

EXAMPLE: 9 inches diameter x 9 = 81 x 3.1416 = 254.4696.

5. RULE to find the cubic inches in a ball. Multiply tube of the diameter by .5236 the answer equals its solid contents.

EXAMPLE: Ball 3 inches in diameter; 3 x 3 = 9; 9 x 3 = 27 x .5236 = 14.1372 solid contents.

RULE to find pressure on the crown sheet of a hanging fire-box boiler. Multiply the width by the length in inches, then multiply by steam gauge pressure and divide by 2.

EXAMPLE:

Crown sheet 46 x 33 in.
Pressure 85 lb.
Iron ½ in.

```
      46
      33
    ----
    1518
      85
    ----
```

If iron is ¼ in. div. by 4. 2)129030

If iron is ⅝ in. div. by 2.66 2000)64515 lbs. pressure.

32.257 tons "

RULE to find how much water a boiler will contain. For a 2-flue boiler, ⅔ full of water, find ⅔ of the area of the boiler in inches inside; multiply by length in inches; then find the area of flues, thickness of iron added; then multiply by 2, if two flues; multiply by length in inches, subtract area of flues from ⅔ contents and divide by 231 (number cubic inches in a standard gallon); the answer will be the number of U. S. gallons.

EXAMPLE:

```
Boiler 48 inches.           48
Two flues, 16 in. each.     48
                          ─────
Length 20 feet.            2304
        16                .7854
                        ─────────
        16              3)1809.5616  area of Boiler.
      ─────               ─────────
       256               603.1872   One-third of area.
      .7854                    2
    ─────────             ─────────
    201.0624             1206.3744  Two-thirds of area.
          2                   240   Length in inches.
    ─────────             ─────────
    402.1248             289529.8560
         240              96509.9520 Sub. Area of flues.
    ─────────             ──────────
   96509.9520           231)193019.9040
                           ──────────
                           835.5940  No. of Gallons.
```

RULE to find the amount of water required, when the average pounds of coal used per hour is known. Divide the coal by 7.5; the answer will be cubic feet; then multiply by 7.5, and that gives the number of U. S. standard gallons.

EXAMPLE:

```
117 lbs. of coal used per hour, 7.5)117.0
                                   ──────
                                     15.6
                                      7.5
                                   ──────
                                    117.0 = 117 gals.
```

Ques. How many cubic feet in 1 lb. of air? Ans. 13.817 cubic feet.

Ques. How much air does it take to consume 1 pound of coal? Ans. It takes 18 pounds, or 248.706 cubic feet.

Ques. How can you tell the water contents of a tank? Ans. If the tank was large at the bottom and narrow at the top, lay the tank off in 10 parts from top to bottom, then take the diameter 4-10 from the large end of the tank, square it, then multiply by .7854; that gives the area; then multiply quotient by full depth of tank and divide by 1728, which gives the number of cubic feet; multiply answer by 7.5, and the number of U. S. gallons will be given. The example must be done in inches; 1728 is the number of inches in a cubic foot; and 7.5 is the number of gallons in a cubic foot.

EXAMPLE:

Tank 2 feet diam.   24 inches diameter.
Tank 3 feet deep,   24 " "

```
            576
            .7854
           ---------
           452.3904  area in inches.
               36   inches deep.
         ----------
    1728)16286.0544
         ----------
            9.4248  cubic feet.
            7.5     No. gals. in a cub. ft.
         ----------
           70.86000 U. S. gals. in tank.
```

RULE for chimneys. Chimneys should be round instead of square, to insure a good draft. The opening should be one-fifth larger than the area of the flues or tubes combined; if less, the draft will not be free. The opening from the bottom should increase in size to the top, and be smooth inside.

Ques. How could the area of any chimney be known for marine or stationary boilers? Ans. To find the area of chimney required for any boiler, multiply the nominal horse-power of the boiler by 112, and divide the product by the square foot of the height of the chimney in feet. The answer will be the required area in square inches. For marine boilers the rule is to allow fourteen square inches of chimney area for each nominal horse-power. In stationary boilers the chimney area should be one-fifth greater than the combined area of all the tubes or flues.

RULE for making good babbitt metal and solders for high and low speed, in parts.

COMPOSITION OF SOLDERS.

Fine solder is an alloy of two parts of Block Tin and one part of lead. Glazing Solder is equal parts of Block Tin and Lead. Plumbing Solder, one part block Tin, two parts Lead.

SOLDERING FLUID :—Take 2 oz. muriatic acid, add zinc till bubbles cease to rise, add ½ teaspoonful of sal-ammoniac.

HIGH SPEED.	COMMON.	MEDIUM.
Martin's Nickel.	10 Copper............	12 Copper........... 60
Copper............	16 Antimony.......	4 Antimony....... 25
Antimony........	4 Tin..............	84 Tin.... 15
Tin.....	70	
	100	100	100

RULE for babbitting a box. Nearly every engineer has his own way; but the best and quickest way is to chip out all the old babbitt in the cap and box, then put the journal or shaft that is to run in the box in its place; put enough liners in between the shaft or journal and edge of box until level, square and in line, put thick putty around the shaft and against the box, so the babbitt can not run out; then heat the babbitt until it runs free, and pour accordingly; the cap is then bolted in its place upon $\frac{1}{16}$ inch thick liner, and putty placed as before; then pour metal through the oil holes which will have to be drilled out afterwards.

TO AVOID BURNING BABBITT METAL:—When melting it for journal bearings, cut part of it into small pieces to cover the bottom of ladle. The block will then be heated evenly instead of being raised to a high temperature at the points of contact with the ladle, which will often become heated to redness before the bulk of the babbitt melts.

RULE to determine the capacity of any size pump, single or double action. Multiply the area of the water piston-head face or plunger in inches, by its stroke in inches, which gives the number of cubic inches per single stroke; the answer divided by 231 (the cubic inches in a gallon) will give the number of standard gallons per single stroke. But remember, all pumps throw less water than their capacity, which depends upon the conditon and quality of the pump. This loss arises from the rise and fall of the valves; from a bad fit or leakage, and in some cases from there being too much space between the valves, piston or plunger. The higher the valves have to rise to give the proper opening, the less work the pump will perform.

FRICTION OF WATER IN PIPES.

Loss in Pounds Pressure per Square Inch by Friction for each One Hundred Feet of Straight Smooth Pipe.

Gals. per Min.	SIZE OF PIPES.										
	1 in.	1½	2	2½	3	4	6	8	10	12	14
5	0.84	0.12	....	....	....	....	....	....	....	....	....
10	3.16	0.47	0.12	....	....	....	....	....	....	....	....
15	6.98	0.97	0.27	....	....	....	....	....	....	....	....
20	12.3	1.66	0.42	....	....	....	....	....	....	....	....
25	19.0	2.62	0.67	0.21	....	....	....	....	....	....	....
30	27.5	3.75	0.91	0.30	....	....	....	....	....	....	....
35	37.0	5.05	1.25	0.42	0.14	....	....	....	....	....	....
40	48.	6.52	1.60	9.51	0.17	....	....	....	....	....	....
45	....	8.15	2.01	0.61	0.27	....	....	....	....	....	....
50	....	10.0	2.44	0.81	0.35	0.09	....	....	....	....	....
75	....	22.4	5.32	1.80	0.74	0.21	....	....	....	....	....
100	....	39.	9.45	3.20	1.31	0.33	0.05	....	....	....	....
125	....	48.1	14.9	4.89	1.99	0.51	0.07	....	....	....	....
150	....	....	21.2	7.00	2.85	0.69	0.10	0.02	....	....	....
175	....	....	28.1	9.46	3.85	0.95	0.14	0.03	....	....	....
200	....	....	37.5	12.47	5.02	1.22	0.17	0.04	0.01	....	....
250	....	....	47.7	19.65	7.75	1.89	0.26	0.05	0.02	....	....
300	....	....	....	28.05	11.2	2.66	0.37	0.09	0.03	0.005	....
350	....	....	....	33.41	15.2	3.65	0.50	0.11	0.05	0.007	.004
400	....	....	....	42.96	19.5	4.73	0.65	0.15	0.06	0.01	.005
450	....	....	....	....	25.0	6.01	0.81	0.20	0.08	0.02	.008
500	....	....	....	....	30.8	7.43	0.96	0.25	0.09	0.04	.017
600	....	....	....	....	....	9.54	1.41	0.38	0.14	0.07	.026
700	....	....	....	....	....	14.32	1.89	0.47	0.18	0.08	.034
800	....	....	....	....	....	....	2.38	0.61	0.22	0.09	.045
900	....	....	....	....	....	....	2.60	0.78	0.27	0.11	.055
1000	....	....	....	....	....	....	3.88	1.94	0.32	0.13	.062
1250	....	....	....	....	....	....	....	1.45	0.49	0.20	.091
1500	....	....	....	....	....	....	....	2.09	0.70	0.29	.135
2000	....	....	....	....	....	....	....	....	1.23	0.49	.234
2500	....	....	....	....	....	....	....	....	....	0.77	.362
3000	....	....	....	....	....	....	....	....	....	1.11	.515
3500	....	....	....	....	....	....	....	....	....	....	.597
4000	....	....	....	....	....	....	....	....	....	....	.914

The friction loss is greatly increased by bends or irregularities in the pipe.

TABLE SHOWING GALLONS OF WATER DISCHARGED IN FIRE STREAMS THROUGH 100 FT. OF 2½-INCH RUBBER HOSE WITH GIVEN NOZZLES (SMOOTH).

Diam. of Nozzle.	Pressure at Nozzle.	Gallons per min.	Horizontal Stream.	Vertical Stream.	Diam. of Nozzle.	Pressure at Nozzle.	Gallons per minute.	Horizontal Stream.	Vertical Stream.
1	30	134	90	62	1⅛	70	259	163	125
1	40	155	109	76	1⅛	80	277	175	137
1	50	173	126	94	1⅛	90	294	186	148
1	60	189	142	108	1⅛	100	310	193	157
1	70	205	156	121	1¼	30	210	96	63
1	80	219	168	131	1¼	40	242	118	82
1	90	232	178	140	1¼	50	271	138	99
1	100	245	186	148	1¼	60	297	156	115
1⅛	30	170	93	63	1¼	70	320	172	129
1⅛	40	196	113	81	1¼	80	342	186	142
1⅛	50	219	132	97	1¼	90	363	198	154
1⅛	60	240	148	112	1¼	100	383	207	164

Ques. Will a boiler 60 inches in diameter, ⅜ inch iron, stand as much pressure as a boiler 48 inch diameter ⅜ inch iron? Ans. No.

Ques. Why? Ans. Because the pressure in the boiler has more surface, and will not allow it. It is the the same as a long bar and a short bar of the same thickness; it takes less strain to break the long one than the short one.

RULE for finding safe working pressure of steam boilers Always use .56 for single riveted and .70 for double riveted side seams. A radius means ½ the diameter and ⅛ of tensile strength is safe load. U. S. Standard is ⅙.

Multiply the thickness of iron by single or double rivets, then multiply by the safe load, divide by internal radius, and the answer will be the safe working pressure.

EXAMPLE:

Diam. 42 in.	.1875 thickness of iron.	
Iron ³⁄₁₆ in.	.70 double riveted.	

```
Double riveted            .131250
50,000 lbs. tensile strength   10000    2)42
                                        ────
                   20.8125)13125000.00    21  outside radius.
Safe working pressure,        63.06     .1875
                                        ─────
                               5     20.8125  inside radius.
                             ─────
Bursting pressure,           315.30
```

RULE to find the aggregate strain caused by the pressure of steam on the shells of boilers. Multiply the circumference in inches by the length in inches; multiply this answer by the pressure in pounds. The result will be the pressure on the shell of boiler, and divide by 2000, which gives the tons.

EXAMPLE:

Diam. of boiler 48 inches, circumference 150.7968, length 20 feet, or 240 inches, pressure of steam 120 lbs, 150.-7968 x 240 x 120 = 4342947.8400 lbs., divided by 2000 = 2171½ tons strain.

RULE to find the number of feet of 1 inch pipe required to heat any size room with steam. For direct radiation 1 lineal foot (straight foot) to 25 cubic feet of space. For indirect radiation, 1 lineal foot to 15 cubic feet of space. Note, all pipe is measured inside for size

EXAMPLE.

Room 18 x 18 x 18 to be heated with 1 inch pipe. Direct radiation. All circulating must be done in inches, and divided by 1728 to find the cubic feet.

$$\begin{array}{r}216\\216\\\hline 46656\\216\\\hline\end{array}$$

1728)10077696 cubic inches.

25)5832 cubic feet.

Lineal 233$\frac{7}{25}$ feet of 1 inch pipe.

One cubic foot of boiler is required for every 1500 cubic feet of space to be warmed. One horse power of boiler is enough for 40,000 cubic feet of space.

RULE to find the horse-power of boilers. Always find the number of square inches and divide by 144, which gives the square feet of heating surface, and divide by 15 square feet for flue boilers, 12 sq. feet for tubular, and 7 sq. feet for cylinder boilers, which is an average allowance for one horse power of boilers; divide the H. P. by 2, you will have the proper grate surface, and allow ½ sq. inch of safety valve to each square foot of grate surface. Generally, from ½ to ¾ of a square foot of grate surface is allowed to each horse power of a boiler.

The H. P. of a boiler is to some extent a misnomer. It is the engine which furnishes the H. P.; the boiler furnishes the steam to make it. A high-grade compound condensing engine will develop a H. P. with fifteen pounds of steam per hour; an ordinary non-condensing mill engine requires thirty. The engines of Watt's time required sixty or more. A boiler which would furnish 3000 lbs. of steam per hour would with the last class of engine, produce fifty H. P.; with a non-condensing engine, one hundred H. P.; and with a compound engine, two-hundred H. P. Watt, basing the estimate upon the consumption of his engines, established the rule of a cubic foot of water per H. P. per hour, but the accepted standard at present, based upon the consumption of the average non-condensing mill engine, is thirty pounds of water from feed water at 100° F. into steam at seventy pounds gauge pressure; equivalent to thirty-four and a half pounds of water evaporated from an atd 212° F.

A common custom among boiler makers has led to the rating of boilers by horse-powers, and while no special harm has come of this, yet it is evidently a misnomer as applied to a boiler, separate and apart from an engine, for the simple reason that a boiler of a capacity suited to a 100 horse power Corliss engine, would only produce about 60 H. P. if attached to a slide valve engine cutting off at four-fifths of its stroke. In the first instance, 3 gallons of water evaporated per hour, produce one horse-power, and in the second, 5 gallons are necessary. To add to the confusion produced by these differences we have another unsettled quantity in the rating of boiler power; namely, the amount of heating surface necessary to a horse-power. The range of difference varies from 10 to 15 square feet, and practice shows that either may be right. For the amount of draft, or intensity of the heat in the furnace will determine this, as, for example, the heat is so great in a locomotive fire-box when the exhaust is strong, that 8 square feet or less of that surface produces a horse-power, and the consumption of coal may run as high as 50 pounds to a square foot of grate surface per hour, while in many furnaces it is reduced to 10 lbs.

From so wide a range in the amount of heating surface and the difference in the consumption of fuel by forced draft, it is evident that when boiler capacity is increased in this way it is done at the expense of fuel.

If we are to make this matter of economy a consideration in the production of steam power, then the true measure of efficiency in a boiler must be its ability to evaporate the greatest amount of water with the least amount of coal or other fuel. The very best results attained in common practice is 10 pounds of water evaporated with 1 lb. of coal, but 8 lbs. is usually satisfactory.

All these varying conditions get us into a dilemma when we come to propose a standard unit for boiler capacity, and as has been said of the jury system, we should let it alone until we can propose something better than horse-powers. We think good may result from drawing a well defined line between the source of power, the boiler, and the means of application of this power, the steam engine. They are too often treated as a whole, and as a result good or bad performance cannot be accurately located.

Appliances are now being introduced for determining the efficiency of boilers, but much remains to be done to simplify them so as to be of general utility.

RULE to find the horse-power generated in any kind of boiler when running. First, notice how long it will take to evaporate one inch of water in the glass gauge, divide this into 60, which gives the number of inches evaporated in one hour; second, multiply the average diameter where evaporation took place by the length of the boiler in inches; this multiplied by the number of inches evaporated, and the answer divided by 1728 gives the cubic feet of water evaporated in one hour.

As a rule, 1 cubic foot of water evaporated is generally allowed for one horse-power; also the capacity of a pump or injector for any boiler should deliver one cubic foot of water each horse power per hour, and an engine uses one-third of a cubic foot of water per horse power.

EXAMPLE:

Length of boiler 216 inches.   216
Average diam. 40 inches.   40

One inch evaporated in 15 min.   8640   15)60

                                    4   4

                            1728)34560(20 horse power.

Weight of Sq. Superficial Foot of Boiler Plate when Thickness is Known.

THICKNESS.		WEIGHT.	THICKNESS.		WEIGHT.
Inches.	Dec.	lbs.	Inches.	Dec.	lbs.
$\frac{1}{32}$ =	.03125	1.25	$\frac{5}{16}$ =	.3125	12.58
$\frac{1}{16}$ =	.0625	2.519	$\frac{3}{8}$ =	.375	15.10
$\frac{3}{32}$ =	.0937	3.788	$\frac{7}{16}$ =	.4375	17.65
$\frac{1}{8}$ =	.125	5.054	$\frac{1}{2}$ =	.5	20.20
$\frac{5}{32}$ =	.1562	6.305	$\frac{9}{16}$ =	.5625	22.76
$\frac{3}{16}$ =	.1875	7.578	$\frac{5}{8}$ =	.625	25.16
$\frac{7}{32}$ =	.2187	8.19	$\frac{3}{4}$ =	.75	30.20
$\frac{1}{4}$ =	.25	10.09	$\frac{7}{8}$ =	.875	35.30
$\frac{9}{32}$ =	.2812	11.38	1 =	.1	40.40

Ques. Explain how the above fractional parts of whole numbers are made to read as decimals—take $\frac{3}{16}$ of an in. for an example? Ans. To do this take 100 as a whole number; divide 16 into 100 = 6¼, reads .625 = $\frac{1}{16}$ of 100. $\frac{3}{16}$ would read, 3 x .625 = .1875. This principle answers for all the rest.

WEIGHT OF A CUBIC FOOT OF EARTH, STONE, METAL, &c.

Article	Lbs.	Article	Lbs.
Alcohol	49	Lead, cast	709
Ash Wood	53	Lead, rolled	711
Bay Wood	51	Milk	64
Brass, gun metal	543	Maple	47
Brandy	58	Mortar	110
Beer	65	Mud	102
Blood	66	Marble, Italian	169
Brick, common	102	Marble, Vermont	165
Cork	15	Mahogany	66
Cedar	35	Oak, Canadian	54
Copper, cast	547	Oak, live, seasoned	67
Copper, plates	543	Oak, white, dry	54
Clay	120	Oil, linseed	59
Coal, Lehigh	56	Pine, yellow	34
Coal, Lackawanna	50	Pine, white	34
Cider	64	Pine, red	37
Chestnut	38	Pine, well seasoned	30
Ebony	83	Platina	1,219
Earth, loose	94	Red Hickory	52
Glass, Window	165	Silver	625¾
Gold	1,203⅔	Steel, plates	487½
Hickory, pig-nut	49	Steel, soft	489
Hickory, shell-bark	43	Stone, common, about	158
Hay, bale	9	Sand, wet, about	128
Hay, pressed	25	Spruce	31
Honey	90	Tin	455
Iron, cast	450	Tar	63
Iron, plates	481	Vinegar	67
Iron, wrought bars	486	Water, salt	64
Ice	57½	Water, rain	62
Lignum Vitæ Wood	83	Willow	36
Logwood	57	Zinc, cast	428

RULE for safety valves. To find the distance, ball should be placed on lever, when the weight is known, or the distance is known and the weight is not known. Multiply the pressure required by area of valve, multiply the pressure required by area of valve, multiply the answer by the fulcrum; subtract the weight of the lever, valve and stem, and divide by the weight of ball for distance, or divide by distance for weight of ball with the same example as follows:

EXAMPLE:

Weight of ball,	60 lbs.	100 lbs. pressure.
Pressure,	100 "	3 area of valve.
Wt. of L. V. & steam,	30 "	300
Fulcrum,	4 inch,	4 fulcrum.
Area of valve,	3 "	1200
		30 wt. of L. V. & st.

60)1170

19½ inch ball should be hung on lever.

The mean effective weight of valve, lever, and stem is found by connecting the lever at fulcrum, tie the valve-stem to lever with a string, attach a spring scale to lever immediately over valve, and raise until the valve is clear of its seat which will give the mean effective weight of lever, valve and stem.

RULE for figuring the safety valve and to know the pressure when the area of valve, the weight of lever, valve and stem, the distance the fulcrum is from valve, and weight of ball is known.

Divide fulcrum into length of lever, multiply answer by weight of ball, add weight of lever, valve and stem, divide by area of valve. Answer will be steam pressure.

Weight of ball,	50 lbs. 2.25	4)20
Wt. of L. V. & stem,	30 lbs. 2.25	5
Fulcrum,	4 in. 5.0625	50
Diam. of valve,	2¼ in. .7854	250
Length of lever,	20 in. 3.9760875 area.	30

Add as many ciphers to the dividend as there are decimals in the divisor, and divide as whole numbers.

3.9)280.0

lbs. press. 71.34/39

To measure or mark off the lever, you measure the fulcrum and make notches the same distance as fulcrum; if fulcrum is 4 inches, each notch must be 4 inches apart.

Ques. What is meant by a fulcrum? Ans. The distance valve stem is from where the lever is connected.

SCREW CUTTING LATHE.

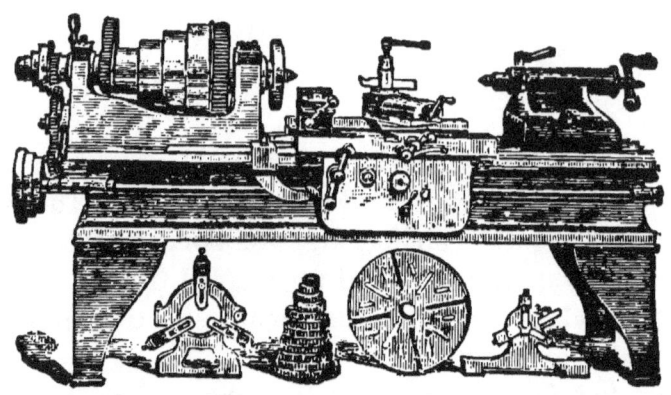

FOR CUTTING ANY SIZE THREAD.

POINTS FOR MACHINISTS.

RULE TO GEAR A LATHE FOR SCREW-CUTTING.—Every screw cutting lathe contains a long screw called the lead screw, which feeds the carriage of the lathe while cutting screws; upon the end of this screw is placed a gear to which is transmitted motion from another gear placed on the end of the spindle; these gears each contain a different number of teeth, for the purpose of cutting different threads, and the threads are cut a certain number to the inch, varying from one to fifty. Therefore, to find the proper gears to cut a certain number of threads to the inch, you will first multiply the number of threads you desire to cut to the inch by any small number, 4 for instance, and this will give you the proper gear to put on the lead screw. Then with the same number, 4, multiply the number of threads to the inch in the lead screw, and this will give you the proper gear to put on the spindle. For example, if you want to cut 12 to the inch, multiply 12 by 4, and it will give you 48. Put this gear on the lead screw, then with the same number 4, multiply the number of threads to the inch in the lead screw. If it is 5, for instance, it will give you 20; put this on the spindle and your lathe is geared. If the lead screw is 4, 5, 6, 7 or 8, the same rule holds good. Always multiply the number of threads to be cut first.

Some—indeed, most small lathes—are now made with a stud geared into the spindle, which stud only runs half as fast as the spindle, and in finding the gears for these lathes you will first multiply the number of threads to be cut, as before, and then multiply the number of threads on the lead screw as double the number it is. For instance, if you want to cut 10 to the inch, multiply by 4, and you get 40; put this on the lead screw, then, if your lead screw is 5 to the inch, you call it 10, and multiply by 4, and it will give you 40. Put this on your stud and your lathe is geared, ready for cutting.

RULE FOR CUTTING A SCREW IN AN ENGINE LATHE. —In cutting V-thread screws, it is only necessary for you to practice operating the shipper and slide screw-handle of your lathe before cutting. After having done this until you get the motions, you may set the point of the tool as high as the center, and if you keep the tool sharp you will find no difficulty in cutting screws. You must, however, cut very light chips, mere scrapings in finishing, and must take it out of the lathe often, and look at it from both sides very carefully, to see that the threads do not lean like fish scales. After cutting, polish with a stick and some emery and oil.

RULE FOR CUTTING SQUARE THREAD SCREWS.—In cutting square thread screws, it is always necessary to get the depth required with a tool somewhat thinner than one-half the pitch of the thread, after doing this make another tool exactly the pitch of the thread and use it to finish with cutting a slight chip on each side of the groove. After doing this, polish with a pine stick and some emery. Square threads for strength should be cut one-half the depth of their pitch, while square threads for wear, may—and should be—cut three-fourths the depth of their pitch.

RULE FOR MONGREL THREADS.—Mongrel, or half V half square threads, are usually made for great wear, and should be cut the depth of their pitch, and for extraordinary wear they may be cut $1\frac{1}{2}$ the depth of the pitch. The point and the bottom of the grooves should be in width $\frac{1}{4}$ the depth of their pitch. What is meant here by the point of the thread is the outside surface, and the bottom of the groove is the groove between the threads. In cutting these threads, it is proper to use a tool the shape of the thread, and in thickness about $\frac{1}{8}$ less than the thread is when finished. As it is impossible to cut the whole surface, at once, you will cut it in depth about $\frac{1}{16}$ at a time then a chip off the sides of the thread, and continue in this way alternately till you have arrived at the depth required. Make a guage of the size required between the threads and finish by scraping with water. It is usually best to leave such screws as these a little large until after they are cut, and then turn off a light chip to size them; this leaves them true and nice.

RECIPES TO TEMPER TOOLS USED DAILY, SUCH AS CHISELS, TAPS, DIES, REAMERS, TWIST DRILLS, COMMON FLAT DRILLS, AND LATHE TOOLS,—To temper flat, cape or side chisels, and common flat drills, put the tool to be tempered in the fire and heat slowly to a cherry **red** color, about 4 inches from the point. Then take it out and put it in the water, point first, about three or four inches, then draw it back quick about an inch from the point, and leave it so until the water will barely dry on the chisel, then take it out, polish it with a piece of sand stone, and let the heat that is left in the body of the tool force its way toward the point; it will be noticed immediately in the change of color. The color of temper for chisels to cut cast iron should be a dark straw, turning to a blue. The temper of chisels to cut wrought iron or steel should be plunged into water after the dark straw color has disappeared and the blue begins to show itself, and left in the water to cool off. In some cases, where the tool is too cold and the temper will not draw, put the tool in and out of the fire often, until the temper shows itself, then cool immediately. If the temper gets to the point of tool before it is polished, it will have to be heated over again. The above rule answers for lathe, planer and shaper tools as well.

Steel tools are given a diamond-like hardness by German engravers who make them white hot, stick them into sealing wax repeatedly until cold, and then touch them with oil of turpentine.

To temper files very hard. Take water 2 measures—no matter what size—wheat flour ½ measure, and 1 measure of common salt. Directions. Mix into a paste; heat the steel to be hardened enough to coat with the paste by immersing it in the composition—after which heat the tool to a cherry red, and plunge it in cold, soft water. If properly done, the steel will be very hard.

To anneal steel is to heat it and bury it in hot ashes and leave it cool with the ashes. Another quick way is to heat it then let it cool to dark color, then plunge into water.

Taps, dies, reamers and twist drills should be tempered in oil. After being heated to a cherry red all over equally, drop the tool in a bucket of oil (plumb) and leave it there until cold; then take it out and brighten it with emery cloth; be careful not to drop it, because it is brittle and liable to break. To draw the temper of taps, reamers and twist drills, heat a heavy ring red hot and enter the tool centrally in the ring, so the heat will be equal from all sides. The hole in the ring should be about three times the diameter of the tool. An old pulley hub would be about right. The color for reamers, taps and twist drills should be dark straw, turning to a blue near the shank; where the color is changing too fast, drop a little water on it; after the right color is obtained, cool off in water. To draw the temper in dies after being cooled in oil, set them (the threads up) on a piece of red-hot iron and draw temper the same color as taps.

For tempering a spring, heat it cherry red, and put it in oil; after it is cool, take it out and hold it over the fire until the oil burns off; then put the spring in the oil again, then in the fire; do this three times; after the last time plunge it into water and cool off.

THE UNITED STATES GOVERNMENT TEMPERING SECRET and durability to the poorest kind of steel. Siegfried's specification reads as follows: "first heat the steel to a cherry red, in a clean smith's fire, and then cover the steel with common salt, purifying the fire also by throwing in salt. Work the steel in this condition, and while subjected to this treatment, until it is brought into nearly its finished form. Then substitute for the salt a compound composed of the following ingredients and in about the following proportions: One part by weight of each of the following substances; salt, sulphate of copper, sal-ammoniac and sal-soda, together with ½ part by weight of pure saltpeter, said ingredients being pulverized and mixed; alternately heat the steel and treat it by covering with this mixture and hammering it until it is thoroughly refined and brought into its finished form. Then return it to the fire and heat it slowly to a cherry red, and then plunge it into a bath composed of the following ingredients, in substantially the following proportions for the required quantity: of rain water, 1 gal.; alum, sal-soda, sulphate of copper, of each 1½ ozs; saltpeter, 1 oz., and of salt, 6 ozs. These quantities and proportions are stated as being practically the best, but it is manifest that they may be slightly changed without departing from the principles of my invention."

DIRECTIONS FOR JOINING BANDSAWS

The following directions for joining band saws are given by the Defiance Machine Works:

Bevel each end of saw the length of two teeth. Make a good joint. Fasten the saw in brazing clamps with the backs against the shoulder, and wet the joint with solder-water, or with a creamy mixture made by rubbing a lump of borax in about a teaspoonful of water on a slate. Put in the joint a piece of silver solder the full size thereof, and clamp with tongs heated to a light red (not white) heat. As soon as the solder fuses blacken the tongs with water and take them off. Remove the saw, hammer it if necessary, and file down to an even thickness, finishing by draw-filing lengthwise.

THE CORLISS ENGINE.

ADJUSTMENT AND SETTING OF THE CORLISS ENGINE VALVES.

It often happens that engineers, under whose control Corliss engines are placed, are not practically acquainted with the operation of the Corliss valve gear, and are at a loss what to do should the gear need adjustment. By carefully observing the following questions and answers, the desired information will be found.

Ques. Into how many classes are the different types of Corliss valve gear divided? Ans. Into two general classes.

Ques. Which are they? Ans. To the first class belong the crab-claw gear. To the second class belong the half-moon valve gear.

Ques. Which is the more favorable and widely known type now in general use? Ans. The half-moon type.

Ques. Why so? Ans. Because the old style crab-claw steam valve opens toward the center of the cylinder, which obstructs the supply passage and forces the steam to pass over and around the valves. This fault is overcome in the half moon type, as the steam valve opens away from the center of the cylinder, thus leaving a clear and direct passage for the steam into the cylinder.

Ques. Do the two different styles make any difference into the opening of the exhaust valves? Ans. No. The difference in the two classes is simply in the direction of movement of steam valves; the exhaust valves open the same in either class, viz.: away from the center of the cylinder.

Ques. What name has the Corliss valve gear? Ans. It is called a detachable valve gear.

Ques. Why is it called detachable? Ans. Because the steam valves open positively at the proper time by the direct action of the working parts of the engine, and continue to open until the connection with the working parts of the engine are broken by detaching or tripping the hook, by action of the cut-off cams.

Ques. How are the steam valves closed? Ans. When the steam valves are detached they are closed by the action of springs, weights, or more generally vacuum dash pots, thus cutting off the supply of steam.

Ques. How is the detachment or tripping determined? Ans. The time in the stroke at which the tripping takes place is known by the position of the cut off cams, which are moved and controlled by the governor.

Ques. Does the cut-off cams trip the hook always at the same point? Ans. No. The cut-off is determined by the requirements of the load on the engine.

Ques. By what name is this cut-off known? Ans The automatic cut-off.

Ques. How is the theory of the Corliss valve motion easily understood? Ans. The theory is easily understood by considering the four valves as the four parts (or edges) of a common slide valve.

Ques. Why are the four valves of the Corliss engine considered as the four parts (or edges) of the common slide valve? Ans. The working edges of the two steam valves answering as the two steam edges of the slide valve, and the working edges of the two exhaust valves as the exhaust edges of the slide valve.

Ques. The Corliss having four valves, and the common slide valve only one, does it not make any difference in setting? Ans. As far as the setting the principle is the same; the only difference is in the adjustment.

Ques. Why does the adjustment make a difference? Ans. The four working edges of the common slide valve are in one solid valve, so that any change or adjustment of one of the edges interferes with the other three. If one edge is to be changed in reference to the others, it must be done by altering the valve itself. The Corliss valves, on the other hand, are adjustable, each by itself, and any one of the valves may be changed without disturbing the other three.

Ques. Can the adjustment be made while running? Ans. When the engineer is familiar with his engine and knows what changes are necessary, the adjustment may be, and is frequently made without stopping the engine.

Ques. How many edges has a slide valve? Ans. Four—two steam and two exhaust.

Ques. Have the Corliss valves the same number of edges? Ans. No. Each Corlisss valve represents an edge of the common slide valve, viz.: two steam edges, two steam valves, two exhaust edges, two exhaust valves.

Ques. How are the valves connected to the eccentric and worked on the Corliss engines? Ans. With the wrist-plate, carrier arm, rocker arm, and reach rod.

Ques. Is the wrist-plate good for any other purpose? Ans. Yes. It modifies the speed of travel at different parts of the stroke, in relation to each other, and gives a quick and constantly increasing speed when opening the steam valves, and a quick opening and closing of the exhaust valves.

Ques. When do the steam and exhaust valves travel slowest? Ans. When they are closed.

Ques. Can the valves of Corliss engines be adjusted when the reach rod is unhooked from the wrist-plate, so the valves may be properly set, independent of the position of the crank? Ans. Yes.

Ques. Are the Corliss valves easily set? Ans. If the engineer has any knowledge, as he should have, of the ordinary slide valve, and of the effect of "lap and lead," as applied to its workings, and will consider the Corliss valve gear in the light of this knowledge, he will soon master the seeming difficulties in his way and find the Corliss gear to be the simplest, most perfect and most easily adjusted of all valve motions.

Ques. How would you go about setting the Corliss valves? Ans. Begin by taking off the black caps or black heads of all four valve chambers. Guide lines will be found on the ends of the valves and on the ends of the chambers, as follows: On the steam valves, coinciding with the working edges of the valves; on the steam valve chamber, coinciding with the working edges of the steam ports. On the exhaust valves and ports, guide lines are also scribed to set them by. The wrist-plate is centrally between the four valve chambers on the valve gear side of the cylinder. A well defined line will be found on the stand which is bolted to the cylinder, and three lines on the hub of the wrist-plate, which, when they conicide with the line on the stand, show the central position of the wrist-plate and the extremes of its throw or travel. To adjust the valves, first unhook the reach rod connecting wrist-plate with rocker arm and place and hold the wrist-plate in its central position. The connecting rods between steam and exhaust valve arms and wrist-plate are made with right and left hand screw threads on their opposite ends, and provided with jamb nuts, so that by slacking the jamb nuts and turning the rod they can be lengthened or shortened as desired. By means of this adjustment, set the steam valves so that they will have ¼ inch lap for 10 inch diameter of cylinder, and ½ inch lap for 32 inch diameter of cylinder, and for intermediate diameters in proportion.

For the exhaust, set them with 1-16 inch lap for 10 inch bore, and ⅛ inch lap for 32 inch bore on non-condensing engines and nearly double this amount on condensing engines, for good results. Lap on the steam and exhaust valves will be shown by the lines on the valves being nearer the center of the cylinder than the lines on the valve chambers. Having made this adjustment of the valves, the rods connecting the steam valve arms with the dash pots should be adjusted by turning the wrist plate to its extremes of travel and adjusting the rod so that when it is down as far as it will go, the sq. steel block on the valve arm will just clear the shoulder on the hook. If the rod is left too long, the steam valve stem will be likely to be either bent or broken; if too short, the hook will not engage, and consequently the valve will not open. Having adjusted the valves as stated, hook the engine in and with the eccentric loose on the shaft, turn it over and adjust the eccentric rod so that the wrist-plate will have the correct extremes of travel, as indicated by the lines on back of hub of wrist-plate. Then place the crank on either dead center and turn the eccentric in the direction in which the engine is to run to show an opening at the steam valve of from $\frac{1}{32}$ to $\frac{1}{8}$ inch, depending upon the speed the engine is to run. This opening will be shown by the line on the valve being nearer the end of the cylinder than the line on the valve chamber. This opening gives the "lead" or port opening when the engine is on the dead center. The faster the engine is to run the more lead it requires, as a general rule. Having turned the eccentric so as to secure the desired amount of lead, tighten it securely, by means of the set screw, and turn the engine over to the other center, and note if the other steam valve has the same lead. If not, adjust by lengthening or shortening the connecting rod to the wrist-plate as the case may be necessary to do.

If the engine has the half-moon, crab-claw, or other gear which opens the valves toward the center of the cylinder, the manner fo the adjustment will be the same, except that the "lap" on the steam valves will be shown when the line on the steam valve is nearer the end of the cylinder, and the "lead" when this line is nearer the center of the cylinder than the line on the valve chamber. The adjustment of the exhaust valves and the amount of "lap" and "lead" will be the same in either case.

To adjust the rods connecting the cut-off or tripping cams with the governor, have the governor at rest, and the wrist-plate at one extreme of its travel. Then adjust the rod connecting with the cut-off cam on the opposite steam valve so that the cams will clear the steel on the tail of the hook about $\frac{1}{32}$ inch. Turn the wrist-plate to the opposite extreme of travel and adjust the cam for the other valve in the same manner. To equalize the cut-off and test its correctness, hook the engine in and block the governor up about $1\frac{1}{4}$ inch, which will bring it to its average position when running. Then turn the engine slowly, in the direction in which it is to run, and note the distance the cross-head has traveled from its extreme position at dead center when the cut-off cam trips or detaches the steam valve. Continue to turn the engine beyond the other dead center and note the distance of cross-heads from its extreme of travel when the valve drops. If the distance is the same as when the other valve dropped the cut-off is equal. If not, adjust either one or the other of the rods until the distances are the same.

By following these directions the engine will do good work, but to know just what it is doing the engineer should use the indicator often. No engine room is complete without a good indicator, and no engineer can be well posted as to what his engine is doing and keep it in its best possible condition for good work without having an indicator and using it often.

THE DYNAMO.

Ques. What is a Dynamo? Ans. A Dynamo is a machine in which Electricity is gathered and forced out through wires for lighting, electro-painting, etc.

Ques. What does a Dynamo consist of? Ans. A Dynamo consists of a field, frame, armature, commutator, brushes, brush holders, pins for the brush holder, and a quadrant.

Ques. What is meant by a field? Ans. It means the magnets connected to the frame with bolts.

Ques. What are magnets? Ans. Magnets are iron cores, wound with insulated wire. These magnets are called electro-magnets because they become magnetic only when a current passes through the wire.

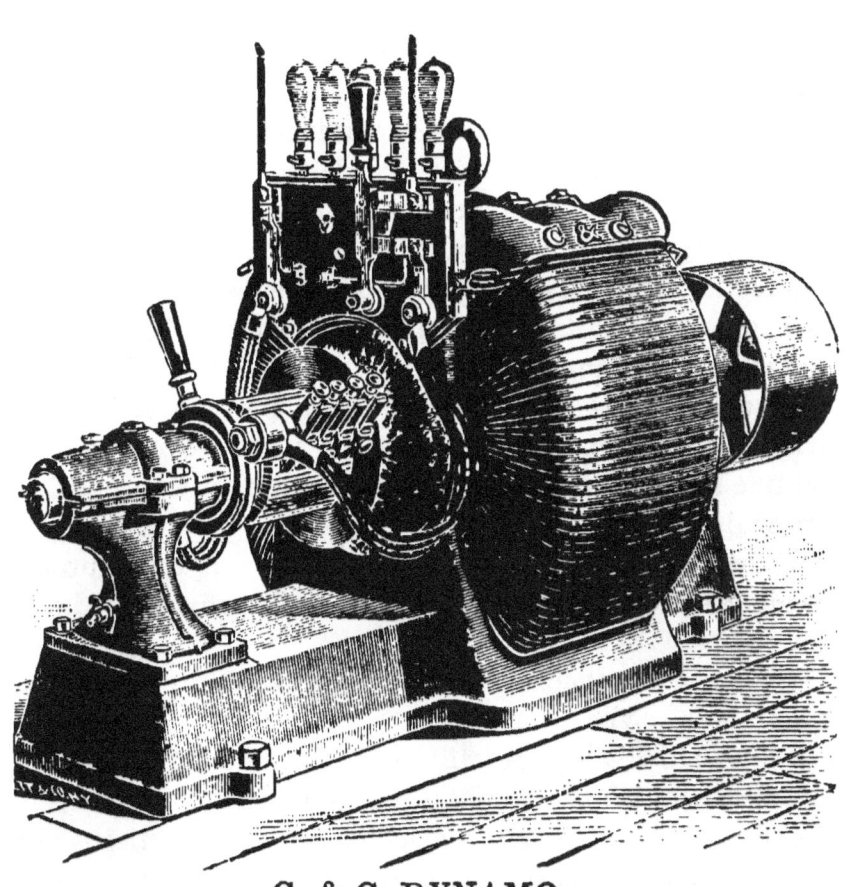

C. & C. DYNAMO.
McDOUGALL & CUMMINGS,
CHICAGO.

Ques. How is the current generated? Ans. By the rotary motion of the armature between the poles of the magnet.

Ques. What does an armature consist of? Ans. It consists of a steel or iron shaft, around which insulated wire is wound, the shaft having a 6 or 8 inch bearing at each end.

Ques. How is this current conducted to the lamps? Ans. By means of brushes made out of copper strips of wires about 6 or 8 inches long, soldered together at one end and held on the commutator by means of brush holders made out of brass. These holders are on long pins, the pins are nutted to a quadrant and the quadrant is fastened to a frame.

Ques. How many brushes are there generally, and where are they? Ans. There are 2 and 4 brushes, two on one side of the commutator and two directly opposite, according to size of machine.

Ques. What is a commutator? Ans. A commutator is made out of segments of copper and segments of insulation.

Ques. Can a commutator be taken off when worn out? Ans. Yes.

Ques. How is it generally done? Ans. By taking out the brushes, brush holders, the pins and the armature from the dynamo, then place the two ends of the shaft on wooden horses, mark the wires connecting the armature and commutator by attaching numbered tags (so as to place them, when the new commutator is put on) then disconnect the wires between the commutator and armature and take off the commutator from the shaft.

Ques. How should a dynamo be looked after and run? Ans. See that the machine is clean, journals cool, and that the proper speed is kept up; see that the brushes are directly opposite each other and that the quadrant and brushes are moved around on the commutator according to the number of lights in use.

Ques. How would you know when to move the quadrant? Ans. By the sparking of the brushes on the commutator.

Ques. What mainly causes the dynamo to flash or spark? Ans. The brushes not being directly opposite through the diameter of the commutator, sometimes not enough pressure on the commutator, sometimes the brushes not far enough around on the commutator, also too much brush surface.

Sparking at the brushes. Some styles of dynamos will spark at the brushes in spite of anything the attendant can do to prevent it, but many other styles of dynamos can be run with absolutely no sparks on the commutator. The first point to be attended to is to get your commutator perfectly smooth, or as near it as possible, with the means at your command, for if the commutator is not true you can not prevent it from sparking.

If you have a slide-rest, use it, and get your commutator round and true from end to end. If you have no slide-rest, a 16 in. bastard file will do nearly as well. Take the brushes and brush holders off, so that you may have plenty of room to work. Start the dynamo to turning very slowly. Hold a piece of chalk so near the commutator that it will mark all of the high spots. Move the chalk slowly from end to end of the commutator, so that all high places on the full length will be chalked. Stop the dynamo and amuse yourself filing off those parts that have been marked by the chalk. If you have noticed while the dynamo was turning about how much the commutator was "out," you can easily tell about how much you will have to file away to bring it true. File off all the places that have been marked, and then start up again slowly, and chalk it again. Repeat the chalking and filing until the commutator is round, and of the same size from end to end.

Next get a piece of shingle, thin board, or a piece of lathe even will do, and wrap a sheet of No. oo. *sand-paper* around it—never use emery paper or cloth—start the dynamo at a pretty lively speed, and smooth the commutator down with the sand-paper, holding the flat side against the work. It is not necessary to work it down to *a polished* surface, although it would be well if it were polished. Now that you have your commutator round and smooth—and it must be so smooth that there are none of the marks left on the commutator, for it was trouble that caused them and if any be left they will certainly cause more trouble.

Now that you know your commutator is in good shape, proceed to set your brushes, being certain that the points of opposite brushes are directly opposite through the diameter. The pressure put on the brushes need only be just sufficient to make *good contact*. It is not necessary to have much pressure to preserve good contact. Should the contact be too slight it will make itself known by a peculiar noise that is indescribable, being neither a snap, crack, or pop, and yet might be called by either of these names You may be sure that the noise will call your attention if you are anywhere near, and after you have once noticed it you will easily recognize it the next time. This noise and considerable sparking will always be present when the brushes do not press heavily enough upon the commutator.

If the brushes are not set with the points directly opposite, sparking will result.

If the brushes are set ahead of the neutral line or back of it they will spark.

When setting four brushes on a commutator that requires two brushes side by side, it is sometimes difficult to get all four of them of an equal length, or evenly divided on the commutator, one or more of them will spark more or less. After rocking the brushes back and forth a trifle to find the point of least sparking, you can then tell by the color of the spark whether the brush should be lengthened or shortened. When the spark is of a decidedly greenish color the brush is too short, but if the spark appears to spatter and shows a reddish hue, then you will find that the brush is too long, or it is so worn, that there is too much of it in contact. By the way, you will find fully as much, if not more, trouble arising from having too much brush in contact, than from having too little.

Cutting of Commutator, scratching and eating away of the segments, is mostly due to the brushes having too much surface in contact, and increase of pressure will wear away the commutator, and having too much of the face of the brush in contact will cause an edge of the segments to become eaten away, and if not attended to, they will, in a very short time, become as rough and uneven as a corduroy road.

With the thicker style of brushes we have never found it necessary, even when running at full load, to have more than one-third of the full end surface of the brush in contact with the commutator, and further, we have found that if we allowed the brush to become so worn that even one-half of the end surface bore on the segments it would cause sparking.

To prevent filing the brushes every day (which would be wasteful,) to keep them in the best of order, we found that they could, with great advantage, be turned the other side up and allowed to wear in that way until the surface became to great. This resulted in getting more than twice the amount of work out of a brush than was possible by filing always from one side, or trimming the ends square as often as they became badly worn. If the

THE EDISON DYNAMO.

commutator becomes very hot you will be quite sure to find that your brushes are badly worn.

Flat Spots on the commutator, frequently explained by laying it to sot spots in the copper, we have always found to result from an entirely different cause. When the marks have the appearance of a blow from the pene of a hammer, it will generally be found to be caused by a loosely connected or badly soldered armature wire connection. A spot of this kind continue to grow larger until the cause of it is removed and the commutator dressed down smooth.

At the end of the segments a spark or stream of fire encircling the whole commutator will sometimes be noticed.

This *may* be caused by an accumulation of oil or copper-dust or dirt, that causes a short circuit, but it will generally be found that the insulation is charred or burned through at some place near where the spark is noticed, and if a careful examination of the armature wires are made you will find that a connection is loose or has very poor conductivity. Allowing the commutator to run hot will increase difficulties of this kind.

THE PRINCIPAL OF THE DYNAMO COMPARED WITH THE STEAM PUMP.

We are often asked how can a dynamo be easily understood; the question coming from engineers who have charge of electric lighting plants.

The whole thing may be compared, in its principles, to the working of a steam pump forcing water through a line of pipe of the same extent as the line wires. The dynamo (or pump) forces electricity instead of water. So long as the dynamo or pump works continuously the pipes or wires are filled with a current of water or electricity, flowing in one direction; in other words, a continuous current. Thus we may say: that a certain number of pounds steam pressure is required to overcome the friction of the water in the pipes, so that so many cubic feet or gallons of water shall be delivered per minute, equally true we can say, so many volts are required to overcome the resistance of the wire, so that the current shall be delivered in so many amperes per minute. Hence, to simplify, we may say pounds of steam pressure = volts; the friction=resistance; the pipe=the wire; current = volume of water in motion, and amperes of electricity=gallons of water delivered at the end per minute. Every engineer knows that the larger the pipe the more gallons water per minute, and the less relative friction, so the larger the wire the more current can be carried and the less resistance, relative to the number amperes delivered. The same analogy holds good in the opposite, for the smaller the pipe or wire, the greater the friction or resistance. Every engineer who uses a steam pump or an injector, knows that there is some point to which, if his pipes were reduced in size, nearly or quite all his power (steam pressure) would be absorbed in friction. So electrically, our voltage may be largely consumed or absorbed by too small a wire; in either case—either the water or the electricity—the result of the work done is in both cases uniform and identical, viz: A continuous current, and is the current that has been generally used for the production of light and power. The other current, which is largely employed in the generation of electrical power, viz: the alternating current, differs essentially from that which we have described above, and in fact our analogy to the working of a pump comes to an end. The current from an alternating dynamo, instead of flowing continuously and directly, is simply a vibratory movement, or a "back and forth flow." Here the supremacy of electricity as a power, or rather as a transmittor of power, comes in, for, returning to one pump, should we at each alternate stroke of the pump reverse the direction of flow of the water, the entire power, or nearly all of it, would be absorbed by its weight, and the friction in the pipes. But electricity being without weight, there is of course no loss by reversing its flow; indeed, the possibilities of application to useful service, dependent on the reversals, are of greatest value. To clearly explain the action of the alternating system, we have to consider the requirements under which electricity does the most acceptable work.

Every engineer who is making electric lights knows that the most satisfactory results, *i. e.*, the best light is obtained by using a dynamo and distributing system of as high voltage as possible, in conjunction with a lamp of low voltage. Here, then, we have two actually opposite conditions, which must be harmonized to produce a perfect result in their action, and which are plainly impossible in the continuous current system, which we have explained by the comparison to our pump; because it is evident, to renew the comparison; that, if we are carrying a pressure (steam), and our line of pipes is calculated to deliver a certain amount of water per minute; if we throttle down at the delivery end, so as to deliver only $\frac{1}{10}$ or $\frac{1}{20}$ of the amount, we shall only be able to do so by reducing our pressure relatively, involving a great loss of efficiency, or incur the risk of destruction to the plant at some point.

Hence we are obliged to provide some appliance which shall intervene to convert the high voltage of the dynamo and circuit to the low voltage of the lamps. When such an appliance is used it is known as a converter system, and the use of an alternating current and converter system are mutually dependent on and necessary to each other.

This system can be compared to the engineer's system of steam heating in his building thus: Suppose he carries 75 lbs. boiler pressure, and the steam is carried into the building in one main pipe, and from that is distributed by risers, etc., to the different radiators in the building. It is evident that he has no use for full boiler pressure on the risers and radiators, as, even if they would stand it for a time, it would be no more effective for heating than a reduced pressure; hence, he puts in a reducing valve in the steam main, between the boilers and risers.

So, then, the converter used in connection with an alternating current is exactly an electrical reducing valve, with a high pressure (voltage) on one side, and a low working pressure (voltage) on the other. Thus, by using this converter he may carry any voltage at the dynamo and primary circuit, reducing into the secondary to conform to the amount of current required. Each current continuous or alternate, have especial fields to which they are adapted, and while both are extensively in use each has its peculiar adaption.

Ques. How do you understand the term "volt"?
Ans. The "volt" is a measure of electro—motive force, or original energy. Corresponding to the dynamic term "pressure," but not of power." It is based on the product of one Daniell cell of a battery.

Ques. How do you understand the term "ohm"?
Ans. The "ohm" is the measure of resistance, and compares to the dynamic term of "loss by transmission." It is based on the resistence offered by a copper wire .05 in. diameter, 250 ft. long; or a copper wire, 32 guage, 10 ft. long.

Ques. How do you understand the term "ampere"?
Ans. The "ampere," is the measure for current or what passes; the intensity, it may be called, and is comparable to the dynamic term of "power transmitted," or "effect." It is the residual force of one "volt" after passing through one "ohm" of resistance.

Ques. How do you understand the term "coulomb"?
Ans. The "coulomb" is a measure of current, qualified by time; one ampere acting for one second of time, comparing in nature with the dynamic "foot-pound."

Ques. How do you understand the term "watt"?
Ans. The "watt" is the unit for dynamic effect produced by electro-motive force or current. It equals 44.22 foot-pounds, or 746 h. p.

Ques. How many "coulombs" in a "watt"? Ans. There are 44.22 "coulombs."

Ques. How many "watts" in an electrical h. p.? Ans. There are 746 "watts" in a h. p.

Ques. How many horse power will it take to run a 50 arc light dynamo. Each arc light equaling 45 "volts" and 8 "amperes" giving 1600 candle power to each light?
Ans. Multiply the "voltage" by the "amperes" then the number of lights lit, and divide by electrical h. p. which is 746 "watts." The answer will be the h. p. of engine required.

FORCE OF A THUNDERBOLT.

It has been calculated that the electromotive force of a bolt of lightning is about 3,500,000 volts, the current about 14,000,000 amperes, and the time to be one-twenty-thousandth part of a second. In such a bolt there is an energy of 2,450,000,000 watts, or 3,284,182 horse-power.

PROTECTING BUILDINGS.

There is a popular saying or proverb, that "lightning never strikes twice in the same place," and the casualties that occur from it are so much more rare than those that happen from other causes that a man who takes no precautions to guard against such is not considered negligent by any except those interested in "protecting" his property against damage by lightning. As a matter of fact, the lightning rod pedler has come to be a standard subject of newspaper jokes and considered a sort of harmless fraud by the public. He is bound to meet with some success, however, in his business, because the subject of atmospheric electricity, and thunderstorms is so little understood even by scientists.

It is not at all probable that thunder storms and lightning discharges of the present day differ in the least from those of a hundred or a thousand years ago, but it is very apparent that a greater loss of life and property occurs from them now than ever before, a fact easily accounted for by the increase of population and of property liable to such damage. The enormous increase in the number of newspapers and of facilities for collecting and distributing such items of news naturally also tend to make the impression that such casualties occur oftener than of yore. However infrequent such accidents occur we know that they do happen sometimes and always unexpectedly and very suddenly. It is, therefore, the part of wisdom and prudence to take such precautions as science and experience teach against these contingencies.

THE CAUSE OF THUNDERSTORMS.

It is generally conceded that the evaporation of water from the surface of the earth, most of which contains some kind of mineral salt in solution, is the primary origin of most of the phenomena of atmospheric electricity, and those who do not fully indorse this theory admit that it is the great agency for stirring up the potential energy derived from the sun and distributing it over the universe, whether in thunderstorms or in the incessant quiet changes that are going on in the various forms of forces employed by nature. Vapor, whether invisible in the ultimate divisibility of its component storms or in a partially condensed condition in the form of clouds, is the vehicle in which is stored the sun's potential energy which we call electricity. Hence the common conception of lightning is that it is an electric fluid packed away in the clouds which may at uncertain times be discharged to the earth with destructive energy in the form of "thunder bolts." The lightning rod, or conductor, as the electrians call it, is regarded as having some sort of power to attract these thunder bolts and convey them to the ground like a pipe would carry water. These popular ideas were derived mostly from the eloquent lightning rod pedler and are not only erroneous but lead to mistakes in the location and setting up of lightning rods that sometimes cause fatal results. Atmospheric electricity is just the same as that we use for motive power, heat and light. The thunder cloud is, to all intents, a condenser plate upon which terminates the polarized chain of a circuit, and its action will depend upon the nature of the opposite condensing plate. If this is another cloud at a distance the discharge will take place between them and have little effect upon the earth, except what is called "induction," that will effect telephone. telegraph, and other wires carrying currents of low potential, and people of a peculiar nervous organization. If, however, the earth forms the opposite condensing plate, which often happens, then the discharge will be from the clouds to the earth and sometimes in the opposite direction. In the latter case all bodies, as well as the air between the clouds and the earth are "polarized" and the discharges always occur in the line of least resistance when the tension rises to a degree greater than the resistance of the circuit can sustain. These discharges are very erratic and very slight circumstances will determine their direction, such as a tree, a man, or an animal standing on moist ground, a vein of mineral, a line of piping in the ground, etc., etc.

This brief and crude effort to portray the nature of a discharge of lightning will convey some idea of the principles of lightning rods to avoid the effect. They are not intended to attract or to convey a discharge of lightning from the clouds to the ground but to supercede the condition of polarization and tension in the space to be protected, and if properly made and put up will diminish the likelihood and frequency of the discharges, but it is undoubtedly true that they also invite them by setting up a line of low resistance.

The theory of the lightning rod is that it practically raises the earth's surface to a height that corresponds with the electric relations of the rod and the air, and the protected area is a cone whose base equals the height of the rod, but this theory applies to the rod itself and is greatly affected by the nature of the buildings in the protected area, their form, material and contents. Just what form this protected area assumes when there are buildings within it or to what points it extends no one has ever yet discovered. But whatever the space protected may be, within it the rod (or conductor) lowers the condition of tension, and either nullifies it or transferes it to the space outside of the protected area. When a charged cloud approaches and sets up an "inductive circuit" to the earth the rod conducts the current quietly to the earth and thus lowers the potential above it so that frequently it does not accumulate sufficiently to cause a discharge—that is a lightning flash. And this is the real purpose of lightning rods.

In putting up lightning rods the object should be to connect with earth every portion of the building, and as, in practice this is impossible with any but metal buildings, they should connect every exposed point and particular care should be taken with the chimneys and smoke stacks. Every chimney lined with soot is a fine conductor of electricity, and if there is a fire in it the warm air ascending to the clouds invites a discharge. Nine buildings out of every ten struck by lightning receive the discharge by the chimneys. Every piece of metal in the construction of the building should be connected with the conductors.

For small buildings iron rods are used. If a large number are used and all properly connected to earth $\frac{1}{4}$ inch galvanized telegraph wire will answer every purpose but for a single conductor not less than half inch rod should be used; solid rod is best for it is the mass or weight of metal that conducts and not its surface and a solid rod presents the least surface to rust. Screw or riveted joints will not do. The rods must have continuous metalic connections.

The most important thing in the whole matter of protecting houses from lightning is the earth connections. Every rod must be connected to water or to earth that is saturated with moisture. Water and gas mains are the best connections provided a good metallic connection is made. A well constantly supplied by a stream affords a good earth connection, but the earth connection is sufficient.

PRACTICAL POINTS FOR ENGINEERS.

Steam-pipes, whether for power or for heating, should always pitch downward from the boiler, that the condensed water, etc., may have the same direction as the steam, or otherwise there will be trouble, unless the pipes are either very short or very large.

Globe valves should always be so placed in steam-pipes that their stems are very nearly horizontal, in order to prevent a heavy accumulation of condensed water in the pipes. Wherever a horizontal steam-pipe is reduced in size there should be a drip to avoid filling the larger pipe partially with condensed water.

In order to make a rust joint that will stand heat and cold as well as rough usage, mix ten (10) parts of iron filings and three (3) parts of chloride of lime with enough water to make a paste. Put the mixture on the joint and bolt firmly; in twelve hours it will be set so that the iron will break sooner than the cement.

TO REMOVE RUST FROM STEEL.

Cover the steel for a couple of days with sweet oil; then with finely powdered unslacked lime (known as quick lime,) rub the steel until the rust is removed; then re-oil to prevent further rust.

HOW TO CLEAN BRASS.

Nitric acid, one part; sulphuric acid ½ part. Mix in a jar, swab on and rub with sawdust.

HOW TO CLEAN DIRTY BRASS QUICKLY.

Finely rubbed bichromate of potassa, mixed with twice its bulk of sulphuric acid and an equal quantity of water will clean the dirtiest brass very quickly.

A soft alloy which attaches itself so firmly to the surface of metals, glass and porcelain that it can be employed to solder articles that will not bear a very high temperature, can be made as follows: Copper dust obtained by precipitation from a solution of the sulphate by means of zinc is put in a cast iron or porcelain lined mortar and mixed with strong sulphuric acid, specific gravity 1.85. From 20 to 30 or 36 parts of the copper are taken, according to the hardness desired. To the cake formed of acid and copper there is added, under constant stirring, 70 parts of mercury. When well mixed the amalgum is carefully rinsed with warm water to remove all the acid, and then set aside to cool. In ten or twelve hours it is hard enough to scratch tin. If it is to be used now, it is to be heated so hot that when worked over and brayed in an iron mortar it becomes as soft as wax. In this ductile form it can be spread out on any surface, to which it adheres with great tenacity when it gets cold and hard.

CEMENT FOR SMALL LEAKS IN STEAM BOILERS.

Experiments have shown the following to be effectual for stopping small leaks from the seams of boilers, pipes, etc. Mix equal parts of air-slacked lime and fine sand; and finely powdered litherage in parts equal to both the lime and sand. Keep the powder dry in a bottle or covered box. When wanted to apply, mix as much as needed to a paste, with boiled linseed oil, and apply quickly, as it soon hardens.

CEMENT FOR IRON WORKS.

It is sometimes advisable to fix two pieces of iron, as pipes for water or steam, firmly together as a permanency. Sal-ammoniac, one part by weight; sulphur, two parts; fine iron borings free from oil. The three should be made with water to a conveniently handled paste. The theory of its action is simply union by oxidation. To drive a nail in hard seasoned wood, dip the points in lard. and they can be driven home without difficulty.

Sewing-Machine Oil.

Best paraffine oil, 1 oz; best sperm oil, 1 oz. Mix and use.

Cement Like That on Postage Stamps:

Mix two ozs. of Mextrine, acetic acid, ½ oz., water, 2½ ozs. After mixture is made, add ½ oz. alcohol.

To Make Tracing Paper

Wet common drawing paper or any other kind, with benzine, the paper becomes transparent immediately, and can be placed over a drawing or picture, to be transferred by tracing with a pencil. This is very valuable.

To Joint Lead Pipes.

Widen out the end of one pipe with a taper wood drift, and scrape it clean outside and inside; scrape the end of the other pipe outside a little tapered, and insert it in the former; then solder the joint with common lead solder by pouring it on with a small ladle and work the solder with a pad made out of 2 or 3 plies of greased bed-tick by holding it under the joint and smoothing it over by working it round making a ball joint, first rubbing a little grease on the scraped parts or joint to be made—thus making it strong.

To Polish Brass.

When the brass is made smooth by turning or filing with a very fine file, it may be rubbed with a smooth fine grain stone, or with charcoal and water When it is quite smooth and free from scratches, it may be polished rotten stone and oil, alcohol, or spirits of turpentine.

To Fill Holes in Castings.

Lead, 9 parts; antimony, 2 parts; and bismuth, 1 part; this is melted and poured in to fill the holes.

To Soften Iron or Steel

Anoint it all over with tallow; heat it in a charcoal fire; then let it cool.

To Distinguish Wrought and Cast Iron from Steel.

File and polish the surfaces, and apply a drop of nitric acid, which is allowed to remain there for one or two minutes, and then washed off with water. The spot will then look a pale ashy gray on wrought iron, a brownish black on steel, a deep black on cast iron. The amount of carbon in iron or steel produces the different colors.

To Case-harden Iron very deep.

Put the iron to be case-hardened in a crucible with cyanide of potash, cover over and heat together, then plunge into water. This process will harden to the depth of three inches.

To Clean Steel and Iron.

Make 1 oz. soft soap and 2 oz. fine emery in a paste; rub it on the article with wash-leather and it will have a brilliant polish.

How to "Figure Out" of the Scrape of Oil-Daubed Sight-Feed Glasses.

Hunt up a plumber friend, and get the use of his gasoline "devil" for ten minutes. Plug one end of a glass tube; then heat about an inch in the middle; when hot, blow into the open end and the tube will quickly bulge itself. Cut off both ends to the right length, and no more trouble will arise from oil-daubed glasses.

Look out for the oily waste, especially if there be any turpentine in it. Keep it in a fireproof receptacle, or better still, burn it up every night. Some dye-stuffs are as bad as oil. The total heat generated by an equal amount of oxidization is identical, whether it proceeds at so slow a rate as to show its effect only in the change of appearance of the article, or so rapidly that the temperature is high enough to consume the substance and ignite the fabric.

A good composition for welding steel is made of one part of salammoniac ten parts borax. The ingredients should be poured together, fused until clear, poured out to cool and finally reduced to powder.

To Find the Height of a Tree or Other Tall Objects.

Take two small sticks of even length, join them together at "C" as shown in the following cut. Place the end "A" on a level with the eye, walk back to such a

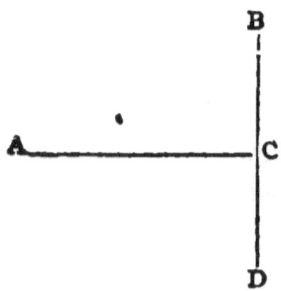

distance from the tree or object that the point "D" may be in a line with the root or base and the point "B" in a line with the top or limb. The distance from the measurer's foot to the root or base of the tree or object, will be equal to the height of the limb or object.

Ques. How can brass and other polished articles be kept from tarnishing? Ans. They can be covered with a thin coat of shellac dissolved in alcohol. The bright work should be warm before applying the coating so it will flow smoothly and dry quickly.

Ques. Give a good recipe of cement that will fasten leather to metal or wood? Ans. Mix a gill of best glue with little water, a teaspoonfull of glycerine, and use warm.

Ques. Give a good black paint for boiler fronts? Ans. Coal tar, ground graphite and turpentine is very durable.

HOW TO RENOVATE BLACK GOODS.

An excellent cleansing fluid, especially useful when men's garments require renovation, is prepared as follows: Dissolve four ounces of white Castile soap shavings in a quart of boiling water. When cold add four ounces of ammonia, two ounces each of ether, alcohol, and glycerine, and a gallon of clear cold water. Mix thoroughly and as it will keep for a long time, bottle and cork tightly for future use. This mixture will cost about eighty cents, and will make eight quarts.

For men's clothing, heavy cloth, etc., dilute a small quantity in an equal amount of water, and following the nap of the goods, sponge the stains with a piece of similar cloth. The grease that gathers upon the collars of coats will immediately disappear, and the undiluted fluid will vanquish the more obstinate spots. When clean, dry with another cloth, and dress the underside with a warm iron. This fluid is also useful when painted walls and wood-work require scouring, a cupful to a pail of warm water being the proper proportions.

RULE FOR CALCULATING SPEED AND SIZES OF PULLEYS.

To find the size of driving pulleys:

Multiply the diameter of the driven by the number of revolutions it shall make, and divide the answer by the revolutions of the driver per minute. The answer will be the diameter of the driver.

To find the diameter of the driven that shall make a given number of revolutions:

Multiply the diameter of the driver by its number of revolutions, and divide the answer by the number of revolutions of the driven. The answer will be the diameter of the driven.

To find the number of revolutions of the driven pulley:

Multiply the diameter of the driver by its number of revolutions, and divide by the diameter of the driven. The answer will be the number of revolutions of the driven.

HOW TO WRITE INSCRIPTIONS ON METALS.

Take ½ lb. of nitric acid and 1 oz. of muriatic acid, mix, shake well together, and it is ready for use. Cover the place you wish to mark with melted beeswax; when cold write your inscription plainly in the wax clear to the metal with a sharp instrument; then apply the mixed acids with a feather, carefully filling each letter. Let it remain from 1 to 10 minutes, according to appearance desired: then throw on water, which stops the process and removes the wax.

RECIPE FOR NICKEL-PLATING WITHOUT A BATTERY.

Take 1½ gills chloride of zinc, 1 gallon of clear water then add enough sulphate of nickel to turn it green, then heat it to a boil in a porcelain vessel. The heating makes the solution cloudy, but does not injure it. Keep the solution next to boiling until the articles to be plated are done say from 30 to 60 minutes then when done polish with chalk first cooling the article in cold water. The articles to be plated should be very clean, to clean articles to be plated use nitric acid 1 part, sulphuric acid ½ part, put in stone jars, then drop in clear water. Strong lye can be used. All handling must be done with a copper wire.

HOW TO PETRIFY WOOD.

Gum salt, rock alum, white vinegar, chalk and pebbles powder of each an equal qoantity. Mix well together, if, after the ebulition is over, you throw into this liquid any wood or porus substance, it will petrify it.

RAILWAY SIGNALS.

One whistle signifies "down brakes."
Two whistles signify "off breaks."
Three whistles signify "back up."
Continued whistles signify "danger."
Rapid short whistles "a cattle alarm."
A sweeping parting of the hands on a level with the eyes, signifies "go ahead."
Downward motion of the hands with extended arms, signifies "stop."
Beckoning motion of one hand, signifies "back."
Red flag waved up the track, signifies "danger."
Red flag stuck up by the roadside, signifies "danger ahead."
Red flag carried on a locomotive, signifies "an engine or train following."
Red flag hoisted at a station is a signal to "stop."
Lanterns at night raised and lowered vertically, is a signal to "start."
Lanterns swung across the track, means "stop."
Lanterns swung in a circle to the left, signifies "back the train."

TIME AND SPEED TABLE.

					Minutes. Seconds. 10ths of sec'nds			
10	miles	per	hour	is................	6.00	to	1	mile.
12	"	"	"	"................	5.00	"	1	"
15	"	"	"	"................	4.00	"	1	"
18	"	"	"	"................	3.20	"	1	"
20	"	"	"	"................	3.00	"	1	"
22	"	"	"	"................	2.43.5	"	1	"
24	"	"	"	"................	2.30	"	1	"
25	"	"	"	"................	2.24	"	1	"
28	"	"	"	"................	2.08.5	"	1	"
30	"	"	"	"................	2.00	"	1	"
34	"	"	"	"................	1.45.6	"	1	"
35	"	"	"	"................	1.42.6	"	1	"
36	"	"	"	"................	1.40	"	1	"
38	"	"	"	"................	1.34.7	"	1	"
40	"	"	"	"................	1.30	"	1	"
41	"	"	"	"................	1.27.7	"	1	"
42	"	"	"	"................	1.25.7	"	1	"
44	"	"	"	"................	1.21.7	"	1	"
46	"	"	"	"................	1.18.2	"	1	"
48	"	"	"	"................	1.15.0	"	1	"
50	"	"	"	"................	1.12.0	"	1	"
52	"	"	"	"................	1.09.4	"	1	"
54	"	"	"	"................	1.06.6	"	1	"
56	"	"	"	"................	1.04,3	"	1	"
58	"	"	"	"................	1.02.2	"	1	"
60	"	"	"	"................	1.00.0	"	1	"

THE STEAMER GREAT EASTERN.

The construction commenced May 1, 1854, and the work of launching her, which lasted from November 3, 1857, to January 31, 1858, cost $300,000, hydraulic pressure being employed. Her extreme length is 680 feet, breadth 82½ feet, and including paddle-boxes, 118 feet; height, 58 feet, or 70 to top of bulwarks. She has eight engines, capable in actual work of 11,000 horse-power, and has besides 20 auxilary engines. She was sold in 1864 for $25,000, and was employed on several occasions with success as a cable-laying vessel. The Great Eastern was sold at public auction October 28, 1885, for $126,000.

USEFUL INFORMATION.

A gallon of water (U. S. Standard) weighs $8\frac{1}{3}$ pounds and contains 231 cubic inches

A cubic foot of water weighs 62½ pounds, and contains 1,728 cubic inches, or 7½ gallons.

Condensing engines require 20 to 25 gallons of water to condense the steam evaporated from one gallon of water.

To find the pressure in pounds per square inch of a column of water, multiply the height of the column in feet by .434. (Approximately, every foot elevation is called equal to one-half pound pressure per square inch.

To find the capacity of a cylinder in gallons. Multiply the area in inches by the length of stroke in inches will give the total number of cubic inches; divide this amount by 231 (which is the cubical contents of a gallon in inches), and the product is the capacity in gallons.

Ordinary speed to run pumps is 100 feet of piston per minute.

To find quantity of water elevated in one minute running at 100 feet of piston per minute: Square the diameter of water cylinder in inches and multiply by 4. Example: capacity of a five inch cylinder is desired: the square of the diameter (5 inches) is 25, which, multiplied by 4, gives 100, which is gallons per minute, (approximately.)

To find the diameter of a pump cylinder to move a given quantity of water per minute (100 feet of piston being the speed), divide the number of gallons by 4, then extract the square root, and the result will be the diameter in inches.

To find the velocity in feet per minute necessary to discharge a given volume of water in a given time, multiply the number of cubic feet of water by 144 and divide the product by the area of the pipe in inches.

To find the area of a required pipe the volume and velocity of water being given, multiply the number of cubic feet of water by 144, and divide the product by the velocity in feet per minute. The area being found, it is easy to get the diameter of the pipe necessary.

The area of the steam piston, multiplied by the steam pressure, gives the total amount of pressure exerted. The area of the water piston, multiplied by the pressure of water per square inch gives the resistance. A margin must be made between the power and resistance, to move the pistons at the required speed; usually reckoned at about 50 per cent.

How to Preserve Eggs.—To each pailful of water, add two pints of fresh slacked lime and one pint of common salt; mix well. Fill your barrel half full with this fluid, put your eggs down in it any time after June, and they will keep two years if desired.

To Kill and get rid of Bedbugs and Moths.—Use either gasoline or benzine. In using these fluids be careful about lights and fires as they are very inflammable.

Egg Stains.—To remove from spoons rub with chloride of sodium.

Hair.—To clean hair, wash well with a mixture of soft water one pint; soda one ounce; cream tartar one-fourth ounce.

Bites and Stings of Insects.—Wash with a solution of water of ammonia.

Bite of Cats.—Apply fat salt pork to the wound for a day or two, or until the poison is all extracted.

Mad Dog Bites.—See a physician at once if possible, or apply caustic potash at once to the wound. Give enough whiskey to cause sleep.

Rattlesnake Bites.—Whiskey is supposed to be the great cure-all. Give enough to cause intoxication.

Burns.—Make a paste of baking soda and water and apply it promptly to the burn. Will check the inflammation and pain.

Screw.—To remove an obdurate screw, apply a red hot iron to the head for a short time, the screw-driver being applied at once while the screw is hot.

Glass Stopper.—To remove a glass stopper from a bottle, warm the neck of the bottle with a warm iron, taking care not to warm the stopper This causes the bottle to expand and loosens the stopper.

Fruit Stains.—To remove the stains of acid fruit from the hands, wash your hands in clear water, dry slightly, and while yet moist strike a match and hold your hands around the flame. The stains will disappear.

Iron Rust.—To remove from muslin or white goods; thoroughly saturate the spots with lemon juice and salt, and expose to the sun. Usually more than one application is required. A good way to prevent its appearance on clothes is when washing to always have them inclosed in a muslin bag when being boiled.

To Stop Vomiting.—Drink freely of hot water, just as hot as can be borne.

Hard Water.—To soften, boil it and expose to the atmosphere. Add a little soda.

Medicine Stains.—To remove from spoons, rub with a rag dipped in sulphuric acid and wash off with soap suds.

To Make A Small Water Filter.—Take a deep flower pot and put a compressed sponge in the hole in the bottom; Over the sponge put a layer of pebbles an inch thick; next an inch of coarse sand; next a layer of charcoal; and at the top another layer of pebbles. The water will filter pure and clear through the aperture into another vessel, however impure previously.

A box having a capacity of one (1) cubic foot will contain ten (10) pounds of cotton waste, packed snugly.

Washed waste is not so economical as cop waste for an engineer's use, though it costs less per pound. Nor is colored so economical as white waste. The best is always the cheapest in the end.

The safe thickness for copper steam pipes may be found by multiplying the diameter in inches by the pressure in pounds and dividing the product by 4000; the quotient is the thickness in parts of an inch. Of course, a trifle more should be added for stiffness and wear.

ANTIDOTES FOR POISON.

In cases where other articles to be used as antidotes are not in the house, give two tablespoonsful of mustard mixed in a pint of warm water. Also give large draughts of warm milk or water mixed with oil, butter or lard. If possible give as follows:

Give milk or white of eggs, in large quantities, for bed bug poison, blue vitriol, corrosive sublimate, lead water, sugar of lead, saltpeter, sulphate of zinc, red precipitate, vermillion.

Give prompt emetic of mustard and salt---tablespoonful of each, follow with sweet oil, butter or milk, for fowler's solution, white precipitate, and arsenic poisoning.

For antimonial wine and tartaric emetic, drink warm water to encourage vomiting. If vomiting does not stop, give a grain of opium in water.

For oil vitriol, aqua fortis, bi-carbonate potassa, muriatic acid, and oxalic acid, take magnesia or soap, dissolved in water, every two minutes.

Drink freely of water with vinegar or lemon juice in it, for caustic soda, caustic potash, volatile alkali.

Give flour and water or glutinous drinks for carbolic acid.

Pour cold water over the head and face, with artificial respiration, galvanic battery, for chloral hydrate, and chloroform.

Prompt emetic, soap or mucilaginous drinks, for carbonate of soda, copperas, and cobalt.

For laudanum, morphine and opium, give strong coffee, followed by ground mustard or grease in warm water to produce vomiting. Keep in motion.

For nitrate of silver, give common salt in water.

For strychnine and tincture nux vomica, give emetic of mustard or sulphate of zinc, aided by warm water.

ANTS.—Sprigs of wintergreen or ground ivy will drive away red ants. Branches of wormwood will drive away black ants. These insects may be kept out of sugar barrels by drawing a wide mark with white chalk around the top near the edge.

BOOTS.—To make leather boots waterproof, saturate them with castor oil. To stop squeaking, drive a peg into the middle of the sole.

CLINKERS.—To remove clinkers from stoves or fire brick, put in about a half a peck of oysters shells on top of a bright fire. It may need repeating but will be effectual.

GREASE SPOTS.—To remove, thoroughly saturate with turpentine, place a soft blotting paper beneath and another on top of the spot, and press it hard. The fat is dissolved and absorbed by the paper.

GILT FRAMES.—To restore and clean gilt frames, gently rub with a sponge moistened in turpentine.

INK STAINS.—To remove, wash carefully with pure water and apply oxalic acid. If the latter changes the dye to a red tinge restore the color with diluted water of ammonia.

PAINT.—Chloroform will remove paint from clothing. When the color of a fabric has been injured by an acid, ammonia is applied to neutralize the same. after which an application of chloroform will in nearly all cases restore the color.

SILVERWARE.—To prevent articles of silverware from tarnishing, first warm them, and then paint them with a thin solution of collodion in alcohol, using a wide, soft brush for the purpose.

STARCH.—To prevent starch from souring when boiled, add a little sulphate of copper.

TO CLEAN FURNITURE.—First rub with cotton waste dipped in boiled linseed oil, then rub clean and dry with a soft cotton flannel cloth. Care must be taken that the oil is all rubbed off.

TO REMOVE SUNBURN AND FRECKLES.—To get off the freckles, to cause the sunburn to disappear, you have got to put on your face and neck, and on your arms, darkened by battling with the waves, a mixture, of two parts of Jamica rum to one of lemon juice; dabble it well on the surface, let it dry, and wash it off in the morning in your hot bath. Besides whitening the skin, which the lemon does, the rum gives it a vigor and makes a rosy flush come to the surface. You will gain no good from this by doing it for one or two nights; keep it up for two weeks at the least, and remember that when your skin has that depressed, wornout look that comes from sitting up too late at night, nothing will invigorate it like a few drops of Jamaica rum put into the water with which you wash your face.

REMEDIES FOR BURNS AND SCALDS.—Every family should have a preparation of flaxseed oil, chalk and vinegar, about the consistency of thick paint, constantly on hand for burns and scalds. The best application in cases of burns and scalds is a mixture of one part of carbolic acid to eight parts of olive oil. Lint or linen rags are to be saturated in the lotion, and spread smoothly over the burned part, which should then be covered with oil silk or gutta percha tissue to exclude air.

STRENGTH OF ICE OF VARIOUS THICKNESS.—Good clear ice two inches thick will bear men to walk on.

Good clear ice four inches thick will bear horses and riders.

Good clear ice six inches thick will bear horses and teams with moderate loads.

Good clear ice eight inches thick will bear horses and teams with very heavy loads.

Good clear ice ten inches thick will sustain a pressure of 1,000 pounds to the square foot.

VALUE OF A TON OF GOLD AND A TON OF SILVER.

The value of a ton of pure gold is $602,799.21.
$1,000,000 gold coin weighs 3,685.8 lbs. avoirdupois.
The value of a ton of silver is $37,704.84.
$1,000,000 silver coin weighs 58,929.9 lbs avoirdupois,

How to Kill Grease Spots Before Painting.—Wash over smoky or greasy parts with saltpeter, or very thin lime white wash. If soap-suds are used, they must be washed off thoroughly, as they prevent the paint from drying hard.

HOW TO MIX PAINTS FOR TINTS.

Red and Black makes........................Brown
Lake and White makes.........................Rose
White and Brown makes....................Chestnut
White, Blue and Lake makes..................Purple
Blue and Lead color makes....................Pearl
White and Carmine makes......................Pink
Indigo and Lamp-Black makes............Silver Gray
White and Lamp-Black makes............Lead Color
Black and Venetian Red makes.............Chocolate
White and Green makes..................Bright Green
Purple and White makes................French White
Light Green and Black makes..............Dark Green
White and Green makes......................Pea Green
White and Emerald Green makes........Brilliant Green
Red and Yellow makes........................Orange
White and Yellow makes..................Straw Color
White, Blue and Black makes..............Pearl Gray
White, Lake and Vermillion makes.........Flesh Color
Umber, White and Venetian Red makes..........Drab
White, Yellow and Venetian Red makes.........Cream
Red, Blue, Black and Red makes................Olive
Yellow, White and a little Venetian Red makes.....Buff

How to Break Glass Any Shape.—File a little notch on the edge of the glass at the point you wish to start the break from; then put a suitably shaped red hot iron upon the notch, and draw, slowly, in the direction you wish. A crack will follow the iron caused by the heat, if not drawn too fast.

How to Drill Glass.—Use a file drill, and keep the point wet with a mixture of camphene and spirits of turpentine. Turpentine can be used alone. The camphene helps the drill to bite. Water will also answer to keep the point of the drill wet.

POLISH FOR FINE FURNITURE.—Linseed oil, and old ale each ½ pint; the white of 1 egg beaten; alcohol, and muriatic acid each 1 oz., mix all together. Directions. Shake well before using and after using keep well corked.

EBONY STAIN FOR SOFT WOOD.—Make a strong decoction of logwood by boiling, and apply three or four times accord ng to shade desired allowing it to dry between applications; then apply solution of acetate of iron. This is made by putting iron filings into good vinegar.

TO WASH AND POLISH SILVERWARE.—One teaspoonful of ammonia to very hot water 1 pt., and wash quickly with a small brush kept for the purpose only, and dry with a clean linen towel; then rub very dry with chamois.

HOW TO CURE DRINKING HABIT.—The cure is simply an orange every morning ½ hour before breakfast, and the keeping away from saloons.

CURE FOR SCIATIA.—Wrap warm flat irons with some woolen fabric, dip in vinegar and apply to painful part two or three times a day, sure cure.

CHOLERA MIXTURE.—Aromatic sulphuric acid, one ounce; paregoric three ounces. Dose. One teaspoonful in four tablespoonfuls of water. This is the simplest and most generally useful combination, and should be kept ready for use in the house, office, store and workshop during a cholera season.

LINIMENT FOR RHEUMATISM.—Alcohol, 1 ounce; oil of mustard, $\frac{1}{16}$ ounce; laudinum 1½ onuce; cod liver oil, 1 pint.

INDEX.

Note.—The index is necessarily very much abbreviated. On account of there being so many points, rules, etc., we found that to index each one would consume too much space. We trust however that any rule required may easily be found by consulting the general heading under which it should come.

	Page.
The Boiler	5
Pumps	19
The Inspirator	26
The Engine	35
Steel Square	49
Valve Motion	54
Horse Power	60
The Indicator	63
Rules	66
The Corliss Engine	88
Valves	88
The Dynamos, etc.	94-109
Practical Points, etc.	110-127

ILLUSTRATIONS.

Flue Brush	11
Pump (common plunger)	20
Steam Pump in detail	21
Duplex Pump in detail	24
Inspirator	27
Boiler Feeder	29
Engine	36
The Governor	39
Lubricator	41-42
Indicator	63
Screw Cutting Lathe	82
The Corliss Engine	88
C. & C. Dynamo	95
Edison Dynamo	100

www.ingramcontent.com/pod-product-compliance
Lightning Source LLC
Chambersburg PA
CBHW020113170426
43199CB00009B/513